Everyone's guide
through the construction process
and a frank inside report from
the files of an architect/builder.

HOW TO GET IT BUILT

Better — Faster — For Less!

Whether you want to add a room,
start that dream house on the hill,
have your restaurant remodeled,
plan a new church sanctuary or
an entire shopping center—

this workbook,
with easy-to-understand examples
and step-by-step guidelines
arranged in the sequence of construction
shows you how to:
obtain drawings, specifications,
bids and all permits;
understand construction financing;
get quality workmanship;
avoid mistakes and foolish shortcuts
and shave building costs 10%, 20%,
or more.

by
Werner R. Hashagen, Architect

Cover and illustrations by the author.

Library of Congress Catalog Card Number: 81-90234

ISBN: 0942514-00-9

SAN: 238-1419

First Printing   1982
Second Printing   1984
Third Printing   1986
Fourth Printing   1987
Fifth Printing   1989

Printed in the United States of America

MY THANKS TO:

Architect David R. Singer for valuable suggestions;
Peggy Luckett who read and typed and read everything one more time;
Sunny Gibbs with HOME magazine of the *Los Angeles Times* for editing;
my wife Ruth for splendid ideas and for support way beyond the call of duty,
and to our clients, who are also our best teachers.

With this publication we give the layperson accurate and enlightening information about the process of building. We do not render legal, accounting, or even technical advice. If such advice is required, the services of a competent professional should be sought.

— Werner R. and Ruth T. Hashagen

Time for a cup of coffee and an inspirational . . .

FOREWORD

A gloomy British proverb states: Only fools build houses—smart people buy them! After 25 years of designing and building, we dedicate these pages to those daredevils, fools or not, who decide to build, or "to get it built":

the investor who envisions a new shopping center,

the church committee that wants a larger sanctuary,

the homemaker who needs an additional bedroom and bath,

the young couple who sees their dream house in a fixer-
upper.

This workbook has been condensed from jobnotes, memos to contractors and clients. The material is organized in the same sequence as the building process:

Part A — 1) Thinking about the project
2) Deciding to build or not to build
3) Getting sketches, drawings, specifications
and permits

Part B — 1) Arranging for financing
2) Lining up contractors
3) Struggling with the budget
4) Signing the construction contract

Part C — Construction

Part D — Appendix

At the end of Parts A, B and C is a checklist that summarizes what you have read. They will help you avoid pitfalls, irrevocable mistakes or foolish shortcuts. The checklists also show you how to shave construction costs to the bone.

Our sympathy is especially with the person who is building for the first time or who, having volunteered to serve on a building committee and, overwhelmed but fearless, has to make all those decisions, decisions, decisions!

Every page assumes that the reader is not an expert in the subject under discussion and that no one will become an expert by reading this book. However, one will become well-acquainted with the in's and out's of the construction process and thereby be better prepared for the adventure that lies ahead. We give you an overview, explain principles and procedures, give examples and conclude the chapters with sensible DOs and DON'Ts.

In this accurate and brief format we hope to familiarize you with building trade lingo and building code legalese, translated into the layperson's terminology. Optimism and a bit of humor are included at no extra charge.

This book does not show you how to design and construct a church, a house or a red-wood deck. **It explains the procedure** of designing and building, that is, how to decide to build or not to build; how to obtain drawings, specifications, permits and bids; how to secure financing; how to find a contractor and work with him during those eventful months of construction, without lying awake at 4:00 every morning.

Our respect and admiration go to the competent architects and engineers, the conscientious contractors and the diminished ranks of dedicated workmen. We thank many of them for the advice they have shared with us.

If you, the reader, have questions, write to us.

For now, good luck! Enjoy the creation, the splendor of action. Be confident and persistent, friendly and hard-nosed. Above all, be patient.

We compiled this guide to help you get your project built successfully—better and faster—and, yes, to save you a bundle.

La Jolla, California   Werner Rolf Hashagen

CONTENTS

PART A: PLANNING AND PERMITS

A-1 SHOULD YOU BUILD FROM SCRATCH, REMODEL OR BUY?

These are the three alternatives: build from scratch, remodel or buy. Many people faced with these choices find it difficult to make up their minds. How do you make a choice? First, consider these factors:

1. What are your needs?
2. What can you afford?
3. What quality do you desire?
4. What location or neighborhood do you prefer?
5. Are there any limitations (legal, budget, neighbors)?
6. What is your time frame?
7. Which adventure are you prepared to cope with?
8. Which will be the best investment?

1. **Decide tentatively what you need:** The number of bedrooms and bathrooms, a den, family room, formal dining room, dining area and breakfast nook or both, sun room, game room, darkroom, expandable areas to be finished later, exterior space to add on. Make a list of **all** your particular needs as well as your desires.

2. **Costs:** Normally it is less expensive to buy than to build. It is also less cumbersome—what you see is what you get! Remodeling is not only the most expensive option (assuming equal quality and square footage), it is also a great inconvenience: you will have to give up the use of a room (or rooms) while under construction, store lumber where you normally park, carpenters constantly ambling through your quarters. On the other hand, remodeling has many advantages: you save the costs of selling and moving, there will be no capital gain problems, nor a big jump in property taxes, and you won't have to start over with landscaping.

3. **Quality:** On a scale of 0 (poor) to 10 (superb), let's assume you aim for a quality level of around 7 or 8. If you don't have that quality in your present house, but can acquire it in a new one, why not move, provided the new house fills your needs in regards to size, location and cost? The same is true for building from scratch: you can specify the quality level of the new building. If you choose to remodel, you can also upgrade the quality of the existing house and build the new portion to equal high specifications.

Needless to say: to protect the value of your building you want to make any addition or renovation blend in with the existing, either by matching what's there or by upgrading the existing to the quality level and the appearance of the new portion.

4. **Location:** Do you like your present neighborhood? Is the landscaping mature and lush, are you close to stores, banks, good schools, the bus stop, the shoemaker, friends and babysitter? Perhaps you despise moving, new subdivisions do not appeal to you, you're wary of the hidden costs of selling. If so, then you should probably stay and remodel. This holds true for a residence, an office, a commercial building or a church. In each case there could be many valid reasons for staying and few, if any, for relocating. But remember that the other side of the coin is: You can improve the house but you cannot improve the neighborhood!

5. **Limitations:** Although certain of these limitations don't meet the eye immediately, they might influence your decision.
   - If you think of remodeling, ask yourself, is there space? Legally, that is. Check with the Zoning Department for the required minimum front, side and rear yards, the allowable maximum coverage, floor area ratio*, etc. When adding a second story to a building, check your title insurance policy for any reference to covenants, conditions or restrictions.
   - Are footings, originally designed to carry the present building, sufficient to support the additional load of another story?
   - If you will spoil your neighbor's view, will you be able to live with his anger for the next 20 years?
   - Read the building code. If the remodeling will be more than a certain percentage in size, normally 50 percent of the existing building, in certain localities the entire building has to be brought up to the requirements of the latest code. It could become an electrician's nightmare!
   - Will you be limited by a certain loan amount needed to buy or build? Maybe you could qualify easily for a remodeling loan, but not for a new-building loan. That is possible if the equity in the existing property is very high, but your borrowing power (the amount of debt you can service) limits the money you could borrow.
   - Do you need the consent of the Coastal Commission or a Home Owners' Association?

6. **How much time do you have?** In one aspect, remodeling saves time. It is likely that you would stay in your house during construction (if you can stand it!) and thereby lose no time in looking for a new house, selling and moving. Building a new structure takes even more time—from 7 to 12 months or longer. Ask anyone who has endured it!

   Time is money! Remember that inflation will continue to increase your costs even if you have plenty of time. The costs of land and construction go up constantly, and are as important to avoid as rising interest rates. Once you have decided to either relocate or remodel, act quickly.

7. **The adventure aspect:** Building or remodeling a house, enlarging an office or restaurant, planning a new sanctuary with fellow church members, are all challenges. Clients should be encouraged to participate as much as possible: measuring or sketching existing conditions, evaluating alternative floor plans that might be submitted, visiting showrooms to select materials or equipment, submitting drawings to government agencies, helping to oversee construction. The more conscientious

---

*For an explanation of unfamiliar terms see the glossary, D-1.

client should enjoy sharing in the work. By doing so, he not only saves on the architect's fee, but gains knowledge about the project that is useful when the client is asked to make decisions or oversee the contractor.

The more time a client can contribute to the building team, from planning through construction, the more the client benefits, not only in actual dollars, but in understanding, satisfaction and most importantly, by exciting memories.

8. **Looking at the project from the investment angle:** For many people this might not be the last, but the first consideration. All right. How do the options rank as an investment? Do you plan to build a Taj Mahal on skid row? Do you want to add a den, a library and a pool room to a one-bedroom house (where you should be adding at least another bedroom)? How could you rate the resale value of the finished product? Get professional opinions from several local realtors before beginning any working drawings.

Can you organize and tabulate your thinking?

To get an overview of your options, fill in the following table. For each CONSIDERATION, rate the OPTIONS from 0 (terrible) to 10 (superb or absolutely no problems). It is like a game.

Use your best educated guesses. (Some ratings will be easier to evaluate after reading the rest of Part A.) Fill in the numbers, then add up each vertical column. The highest sum at the bottom will indicate the best choice.

But first make some notes on those eight considerations:

What are your needs? _____

_____

What can you afford? (Costs, interest rates, etc.)_____

_____

What quality do you desire? _____

_____

Location and Neighborhood: _____

_____

Any limitations? (C.C.&Rs, Zoning Code, soil) _____

_____

What's your time frame? _____

_____

Do you mind the adventure of building? _____

The Investment angle: _____

_____

Now fill in the table on the next page.

5

NOTE: Although we seem to have emphasized residential building in this first chapter - the principles of construction pertain to all types of buildings: residences, apartments, churches, commercial structures, shopping malls, parking garages, skyscrapers, etc.

CHECKLIST

CONSIDERATIONS	Build from scratch	Remodel	Remodel & provide for future finishing or expansion	Buy a new place	Buy another existing house, etc.	Stay put and do nothing
Rate each option for how it can best give you what you **need** in regards to: (a) Size (b) Plan arrangement (c) Appearance						
Considering all **costs** (present property taxes, increased property taxes, selling fees, surveys, soil reports and other building-related expenses, financing, drawings, construction costs, contingencies) rate how well you could afford each option.						
Rate each option for how it would provide the **quality** you desire.						
Rate the **location** (schools, transportation, shopping, neighbors, prestige, landscaping).						
Rate your chances of coping with **limitations** (Is there space—legally, practically? Building Code and CC&R restrictions, Coastal Commission, soil conditions, slopes, Zoning Code restrictions, financing, neighbors' objections). 0 = lots of problems 10 – no problems						
How would each option suit you **timewise**?						
Rate how little (0) or how much (10) you would enjoy the **challenge** of each option.						
Rate each option for its **investment potential.**						
TOTALS						

Now you know—almost!

A-2 STATISTICS

Cost of Houses:
"If present trends continue in the 1980s, only the most affluent will be able to afford one."

—MIT-Harvard Center Report

Investment value of real estate:
U.S. News and World Report stated that among major forms of investments, only real estate (housing, farmland, commercial buildings), very few stocks and some money market funds have kept pace with inflation. The other well-known types of investments, which did not stay ahead of inflation, in decreasing order of performance, are:

Certificates of deposit
Savings bonds
State and local bonds
Corporate bonds
Savings accounts
Most stocks

Some other unconventional investments have performed well in the last 10 years, even better than most real estate. When a friend borrows this book, he may tell you afterwards that this information has absolutely nothing to do with the subject of this guide. But you wanted to know anyway, so here are the investments that beat real estate. According to Salomon Brothers, they rank in the following order, from the highest return to the lowest:

Chinese Ceramics
Rare books
Antique Furniture
Stamps
Coins
Diamonds
Oil
Old masters

Although we believe one should build to satisfy personal needs and not worry constantly about "resale value," consider the following owners' and renters' preferences, as based on surveys by Walker & Lee, a large California Real Estate firm, the San Diego Union and our own experience.

Presently the most preferred types of housing (in decreasing order of preference) are:
>
> 1-story detached single-family residence with some level changes*
> 1-story detached single-family residence (no level changes)*
> Split-level detached single-family residence*
> 2-story detached single-family residence*
> Townhouse
> Condominium apartment
> Rental apartment
>
> *For colder regions add "plus basement."

For all types of housing almost nationwide the most desirable features are:
>
> A 3-bedroom floor plan, approximately 2000 square feet
> oversized, enclosed garage (not a carport)
> Family room
> 2 full bathrooms plus a powder room (l/2 bath)
> Informal eating area, in or near the kitchen
> Formal dining area
> Fireplace (preferred location: the family room)
> Ample storage areas and closets
> Mirrored wardrobes
> Laundry area near the kitchen
> Built-in bookshelves
> Shake or tile roof
> Large rear yard
> A fourth bedroom which can be used as a den or a guest room

Additional optional features liked or preferred by homeowners or renters:
>
> Ceramic tile counter tops (in lieu of laminated plastic)
> Double oven, self-cleaning
> Microwave oven
> Trash compactor
> Food center
> 3-compartment kitchen sink (1 for garbage disposal)
> Water closet separated from pullman and tub area
> Oversized tubs (5 feet 6 inches or 6 feet in lieu of 5 feet)
> Separate tub and shower
> Bathroom fan in addition to window
> Double doors at entry, den, master bedroom
> French doors
> Window seat
> Sloped ceiling (although difficult to heat!)
> Skylight
> Bay window, window seat
> Barbeque cooktop
> Security locks (on wood doors and sliding doors)
> Screw type garage door opener, not chain
> Provision for cable TV
> If there is an intercom or built-in radio system: wire the door chime into it
> Wet bar
> Storage space for bicycles or motorbike

Which rooms should face the backyard:
>
> First choice: family room

Second choice: kitchen
Third choice: master bedroom

Desired energy-conserving features:
Better-than-standard wall and ceiling insulation
Double-glazed windows
Insulation of hot water pipes
Warm-air-circulating fireplace
Attic fan or at least an electrical switch for a future attic fan
Weatherstripping on doors and windows. Air-lock vestibules.
Solar system for hot water heating

Least wanted or appreciated features include:
Sunken conversation pit
Fiberglass tubs and showers
Small-pane windows
Washer and dryer location in garage or bedroom area
Two single garage doors in lieu of one double-size door

Preferences, of course, vary from person to person, from region to region. We don't necessarily agree with all choices and priorities as stated above. Yet, this information can help you specify your own desires as well as provide an insight to majority preferences in coming years. Who knows, maybe single-car garages will be "in" next year!

"Meanwhile, inflation marches on, making it possible for people in all walks of life to live in more expensive neighborhoods without even moving."

CHANGING TIMES
The Kiplinger Magazine

A-3 WHAT IT TAKES TO BUILD

So, you have decided to build. You want either to remodel or build a completely new structure. If "building something" can be compared to a journey into the unknown, would you start traveling around the world or venture into the deserted lagoons of Baja without some idea of what you want to see, how you want to travel, how much you are going to spend? Let's think about it, with pencil and paper. What does it take to build?

FIRST: Take three sheets of lined 8½ by 11-inch paper and make three lists: **Definitely, Maybe** and **Do Not Want.** Tabulate your needs on these lists. Think in terms of rooms, shapes, features, colors, textures and arrangements. Write them on the respective list:

        rooms or features you **definitely** want

        maybe, if it fits into the budget

        what you **don't want**

For instance (and you pick the list): large master bedroom – more storage – powder room – parquet at entry – zoned heating – french doors –greek columns – turkish sauna – fireplace in family room – false beams – real wood beams – wood deck over canyon – swimming pool – connections for future whirlpool – spiral staircase – windowseat.

SECOND: **Gather basic information** about your property, be it a vacant lot or an existing building. We ask our clients to bring to the first meeting as much as they have of the following:

        drawings of lot or existing building

        address of owner and building site (sometimes they are different)

        photographs of building or lot, in color or black and white; simple
          pictures, straight on, sides and rear, and over each corner

        legal description and a copy of the plat map (from the title policy)

        lot size, normally shown on title policy plat

        name and address of legal owner, if different from the client's

        survey or soil information, if already existent

        copy of Covenants, Conditions and Restrictions, if any

        north direction, prevailing wind, view aspects

        If known: building zone, fire zone, location of utilities (water, sewer, electrical, gas, cable TV). For a new residence: get a house number assigned from the water department

and, yes, her or his favorite clippings from *House Beautiful, Sunset Magazine* or *Mechanix Illustrated*—we will look at all of them

THIRD: **Think about a tentative budget.** No need to be on target within $1000; the budget will change anyway. Just think out loud and write down a few round numbers. If you intend to add 1000 square feet to your house, you might assume that you cannot build for much less than $38 per square foot. So, there is $1000 × $38 = $38,000, minimum. Add to the building costs an amount for incidental costs—let's call them building-related expenses:

selling or buying costs (commissions, escrow, legal fees, etc.)
title policies for building loans
financing cost of construction loan; fund control
insurance during construction
drawings, specifications, construction administration
permits
contingencies
setting up a separate construction checking account (check
    printing)
work performed under separate contracts or by owner
beer for the topping out party

Roughly, for a tentative budget, figure about 15 to 20 percent of the building cost. That's about $5,700 to $7,600 to be added to the $38,000. It must be emphasized here that you are not "nailing down the cost" of the project. You are just thinking about a budget.

FOURTH: What else is required of you in order to build successfully? A few old-fashioned intangibles—the more of each the better:

- understanding the construction process
- far-above-average patience
- a spirit for adventure
- an inquisitive mind
- time—lots of it
- a sense of humor
- perseverance
- **enthusiasm**

Before anything gets underway most clients are more concerned with costs than any other aspect of building. Therefore, estimating tentative building costs is the subject of the next chapter.

A-4 BUILDING COST VALUATIONS

Even before drawings are started, one question makes clients wake up at 4:00 in the morning: How much is this project going to cost?

The question will not be completely answered until after drawings and specifications have gone out for bids. There are, however, methods to closely **estimate** the costs of a project earlier, thus allowing the owner to decide whether or not to build, to scale down the project, or to add extra square feet or features, even a few nonessential but desirable options. Some of these options come from the clients "MAYBE" list and when dealing with exotic, dramatic, dazzling or gemütlich features, are referred to among designers as the "Oooh's and Ahh's."

There are at least four types of cost valuations. The first two, which we discuss briefly, are unreliable. In case these methods are suggested to you later on, you will know why to avoid them. The third one, "Square Foot Cost Estimate" is a good method. The fourth, signed and dated firm bids, tells the full story.

Here are the four methods to **estimate** the cost of a project:

1. **The tax assessor's "full value" of a comparable existing building.** Tax valuations are based on market value and take into account the neighborhood, the age of the building and some tax assessor's idiosyncrasies. They don't help much to figure the construction cost of a proposed similar building.

   Don't waste your time.

2. **Building Department "Permit Valuation":** also inaccurate. At best, it indicates the bottom cost of a particular building per square foot.

   The next methods are used by the professionals:

3. **Cost estimated by square footage:** A fast reliable method if done carefully. Experienced architects and contractors know the cost of buildings, either from recently completed projects or from industry reports such as the Dodge Building Cost Services. The **square footage** of a proposed building has to be multiplied by the **square footage cost** of an actually completed project. Good preliminary drawings, with floor plans, sections and elevations and outline specifications are needed to come up with

realistic preliminary estimates. Often a seasoned estimator divides the total square footage of a project into distinctive cost areas. To use a residence as an example, he would not only say, "This plan calls for 2800 sq. ft. at $50.00, that's $140,000 (who builds the garage?) but he would itemize the cost as follows:

Habitable areas (living, dining, bedroom, hall)	2500 sq. ft. @ $42.00*	$105,000
Kitchen and bathrooms	300 sq. ft. @ $75.00	22,500
Garage	500 sq. ft. @ $16.00*	8,000
Driveways and walks	800 sq. ft. @ $ 1.50	1,200
	Preliminary Cost Estimate:	$136,700

*If there is living space over the garage, use for the garage cost the same square footage cost as you use for habitable areas. For balconies, decks, trellises, covered walkways, use one-third of the cost of habitable areas, for unfinished rooms one-half.

$136,700 is the amount the residence could be built for, at the quality level indicated by preliminary drawings and specifications. This **does not** include certain building related costs such as: survey and soil report (if needed), soil compaction or blasting, drawings, specifications, permits, financing and escrow for the construction loan, insurance, bonds, fund control, landscaping, draperies or any unusual construction items not shown on the preliminary drawings. If built under F.H.A. or V.A. regulations, add about $1.00 per square foot.

The "best bid" received by competitive bidding should come within approximately 10 percent of the preliminary cost estimate.

An architect, if using the standard A.I.A. contracts, is obligated to furnish the client with preliminary cost estimates (called "Probable Construction Costs") at various phases, normally once in the design phase and once before bidding.

4. **Firm Bids:** There is a distinction between **negotiated** (or "single") bids and **competitive** bids submitted by several contractors. Both types of bids are based on complete drawings and specifications, including any addenda that might have been issued. Firm bids cannot be obtained if the drawings have not cleared the Building Department and all other governing authorities, such as Coastal Commission, Fire Marshal or Homeowners' Association. The contract should not necessarily be awarded to the **lowest bidder** but, rather, to the **best bidder.** The best bidder is one who submits a low bid, is very experienced, can complete the project on time and is financially stable. A legal note: although the architect helps the owner to select bidders, he cannot and shall not guarantee the contractor's performance in any regard: cost, time, workmanship. The spread in bids may be considerable, difficult to believe from a layperson's view and professionals alike. If the best bid is much too high, there are ways to bring it down. See B-10.

DO:
Use either one or both of the following valuation methods:
  a. **In the planning stage, for a budget figure and for the building loan application:** Preliminary Cost Estimates based on cost-per-square-foot, calculated by an experienced architect or contractor.
  b. **To finalize the building budget and the bank loan:** Firm written bids from qualified contractors, including a contingency allowance.
These two methods will provide the construction cost. To this add all incidental "building-related costs," as close as you can estimate. See A-9

THE CITY OF
SAN DIEGO
BUILDING
INSPECTION
DEPARTMENT
236-6270

AVERAGE COST SCHEDULE FOR ESTIMATED

BUILDING VALUATION

THIRD FLOOR — CITY OPERATIONS BUILDING — 1222 FIRST AVENUE

INFORMATION

21

BULLETIN

Occupancy and Type	Value Per Sq.Ft.
APARTMENT HOUSES:	
*Type I or II F.R.	$71.00
Type V-masonry (or Type III)	55.00
Type V-Wood Frame	47.00
Type I Basement Garage	25.00
BANKS:	
*Type I or II F.R.	100.00
Type III, One-hour	82.00
Type III-N	78.00
Type V, One-hour	71.00
Type V-N	68.00
CHURCHES:	
Type I or II F.R.	66.00
Type III, One-hour	54.00
Type III-N	52.00
Type V, One-hour	49.00
Type V-N	47.00
CONVALESCENT HOSPITALS:	
*Type I or II F.R.	93.00
Type III, One-hour	75.00
Type V, One-hour	68.00
DWELLINGS:*	
Type V-Adobe	79.00
Type V-Masonry	63.00
Type V-Wood Frame	59.00
Additions-Wood Frame	59.00
Basements (non-habitable)	19.00
Solariums	59.00
HOSPITALS:	
*Type I or II F.R.	111.00
Type III, One-hour	104.00
Type V, One-Hour	95.00
HOTELS & MOTELS:	
*Type I or II F.R.	69.00
Type III, One-hour	59.00
Type III-N	56.00
Type V, One-hour	50.00
Type V-N	48.00

Occupancy and Type	Value Per Sq.Ft.
INDUSTRIAL PLANTS:	
Type I or II F.R.	$40.00
Type II, One-hour	27.00
Type II - (Stock)	26.00
Type III, One-hour	29.00
Type III-N	28.00
Tilt-up	19.00
Type V, One-hour	27.00
Type V-N	26.00
MEDICAL OFFICES:	
*Type I or II F.R.	84.00
Type III, One-hour	67.00
Type III-N	64.00
Type V, One-hour	62.00
Type V-N	59.00
OFFICES:	
*Type I or II F.R.	73.00
Type III, One-hour	52.00
Type III-N	49.00
Type V, One-hour	47.00
Type V-N	45.00
PRIVATE GARAGES:	
Wood Frame	14.00
Masonry	18.00
Open Carports	10.00
PUBLIC GARAGES:	
*Type I or II F.R.	33.00
Type II-N	22.00
Type III, One-hour	24.00
Type III-N	22.00
Type V, One-hour	19.00
RESTAURANTS:	
Type III, One-hour	64.00
Type III-N	60.00
Type V, One-hour	56.00
Type V-N	54.00

Occupancy and Type	Value Per Sq.Ft.
SCHOOLS:	
Type I or II F.R.	$80.00
Type III, One-hour	58.00
Type III-N	55.00
Type V, One-hour	53.00
SERVICE STATIONS:	
Type II-N	44.00
Type III, One-hour	47.00
Type V, One-hour	44.00
Canopies	18.00
STORES:	
*Type I or II F.R.	56.00
Type III, One-hour	45.00
Type III-N	43.00
Type V, One-hour	35.00
Type V-N	33.00
THEATERS:	
Type I or II F.R.	77.00
Type III, One-hour	54.00
Type III-N	52.00
Type V, One-hour	48.00
Type V-N	46.00
WAREHOUSES:**	
Type I or II F.R.	34.00
Type II, One-hour	20.00
Type II-N	19.00
Type III, One-hour	23.00
Type III-N	22.00
Type V, One-hour	20.00
Type V-N	19.00

F.R. - Fire-resistive
N - Nonrated

* Add 0.5 percent to the total cost for each story over three.
** Deduct 11 percent for mini-warehouses.
*** For subdivisions with 10 or more single-family dwellings which have plancheck and building permit issuances in groups of 10 or more, the valuation may be decreased by 10 percent.

A typical Building Department's valuation table showing assumed construction costs per square foot for various building types. Assumptions (1987) are on the low side and are used to establish the costs of plan check and permit fees. These tables are updated every year.

The value assumptions are modified for items such as fire sprinklers, air conditioning, pile foundations, fireplaces and other considerations. See next page.

GENERAL ADDITIONS AND MODIFIERS TO BASIC VALUATION

Fire Sprinkler System - $1.50 per square foot of area sprinkled.

Air Conditioning - $2.80 per square foot of floor area for commercial.
$2.40 per square foot of floor area for residential.

Tract Dwellings - 10 or more units - reduce valuation by 10% each unit.

Pile Foundations - increase valuation by $12.00 per lineal foot of pile for cast-in-place concrete piles and $29.00 per lineal foot of pile for steel.

Other dwelling cost modifiers - fireplace: $1,975 each fireplace over two (concrete or masonry)
$1,350 each fireplace over two (prefabricated metal)
Bathroom: $2,250 each bathroom over three

Shell Buildings - the valuation of nonresidential buildings, for which permits are issued without complete interior improvements, will be established as though they were complete store buildings. The valuation for tenant improvements will be determined as the difference between the valuation for a store and the new use which the tenant improvement creates, but not less than $15/sq.ft.

CHARGES TO BE ADDED TO BASIC VALUATION FOR RESIDENTIAL ADDITIONS AND ALTERATIONS

Fireplace
- $1,975.00 each (masonry or concrete)
- $1,350.00 each (prefabricated metal)

ALTERATIONS TO EXISTING STRUCTURES (NO ADDITIONAL FLOOR AREA OR ROOF COVER)

Enclosure of open porch or sunshade, conversion of garage and similar work not affecting floor area - difference in valuation per square foot between existing and new occupancy.

Interior Partition - $27.00 per lineal foot
Install Windows or Sliding Glass Doors - $ 8.50 per square foot of opening
Close Exterior Wall Opening - $ 8.00 per square foot of opening

MISCELLANEOUS

Agricultural Building . . . . $11.00/sq.ft.

Aluminum Siding . . . . $ 3.00/sq.ft.

Antennas. . . . . $2,000.00 each

Awning or Canopy
(Supported by Building)
Aluminum . . . . $12.00/sq.ft.
Canvas. $ 5.00/sq.ft.

Balcony, Stairs and Decks . . . . $ 8.00/sq.ft.

Carport . . . . $ 9.00/sq.ft.

Greenhouse . . . . $ 3.00/sq.ft.

Plastering
Inside . . . . $ 1.50/sq.ft.
Outside . . . . $ 1.50/sq.ft.

Patio
Metal Frame - Cover & Walls . . . . $ 6.00/sq.ft.
Wood Frame - Cover & Walls . . . . $ 5.00/sq.ft.
Masonry Walls. . . . . $ 5.00/sq.ft.
Screen/Plastic Walls . . . . $ 2.00 sq.ft.
Enclosed Patio - Cover & Walls. . . . . . . $ 7.00 sq.ft.

Retaining Wall
Concrete or Masonry . . . . $ 10.00/sq.ft.

Reroofing (1 square = 100 square feet)
Built-up. . . . . $ 76.00 per square
Composition Shingles. . . . . $ 71.00 per square
Asbestos Shingles . . . . $ 168.00 per square
Wood Shingles . . . . $ 168.00 per square
Wood Shakes . . . . $ 168.00 per square
Aluminum Shakes . . . . $ 252.00 per square
Clay Tile . . . . $ 213.00 per square
Concrete Tile . . . . $ 180.00 per square

Roof Structure Replacement . . . . $ 8.00/ sq.ft.

Spa, Hot Tub or Jacuzzi $4,100.00

Saunas (Steam) . . . . $5,100.00 each

Swimming Pool. . . . . $ 21.00/sq.ft. of surface area

Stone and Brick Veneer . . . . $ 5.00 sq.ft.

Fence or Freestanding Wall
Wood/Chain Link . . . . $ 1.10/sq.ft.
Wire . . . . $ 1.10/sq.ft.
Masonry . . . . $ 5.00/sq.ft.
Wrought Iron . . . . $ 3.00/sq.ft.

It is likely you have heard horror stories of how building costs can escalate, driving building committees up the wall and ruining the marriages of the nicest people. What can happen to your carefully calculated building budget is addressed in the next chapter.

A-5 IT ALWAYS COSTS MORE—ESPECIALLY REMODELING

This chapter deals with cost overruns, also called "extra work" or sometimes the "owners-might-as-wells." Let's assume the owner wanted to add a den and powder room and enlarge the living room. He had good drawings and specifications and four eager contractors interested in bidding. Yet, one contractor did not submit a proposal.

The owner received three bids: $29,500

$26,450

$24,500

The $24,500 bidder wanted the work. He had just finished a custom house and his next job was to be a restaurant remodel, but the financing was dragging. He bid as tight as he could—he wanted the job to keep his crew going (often a reason for the best bid).

How could there be a cost overrun on this small job? Especially since the owner remembered a few points he had discussed with the contractor when he signed the contract:

- The contractor thought drawings and specifications were clear and adequate. He had a few questions and they were cleared up.

- The contractor had given a **written** proposal, dated and signed, for $24,500, to be completed in 150 calendar days.

- He understood that the owner would do certain work himself or under a separate contract: final cleaning, window washing, carpet installation, landscaping.

And the owner knew about **building-related costs** that were not part of the contractor's bid and contract: Loan fees, insurance, permit, blueprinting for construction, architect's job visits, filing Notice of Completion.

But during construction the following additional costs developed and were added to the $24,500 contract:

a. When inspecting the footing trenches, the Building Inspector noted soft soil pockets where the den was planned. The contractor had to make deeper footings, remove some soil from the site and add a 6-inch layer of sand . . . . $ 540.00

b. Owner upgraded windows. In lieu of color anodized aluminum, he chose wood windows. (This also added to the painting costs) . . . . $ 480.00

c. When connecting the new bath plumbing to the existing plumbing, some corroded steel pipes had to be replaced with copper piping . . . . $ 415.00

d. Owner instructed painter to repaint entire living room, not just the new addition. (The contractor knew this made sense, but to keep his bid low, had not included this cost.) . . . . $ 245.00

e. Relocate one hose bibb where the remodeling took place . . . . $ 45.00

Total Extra Work: $1,725.00

The owner was surprised that the cost overrun amounted to that much. He should not have been. Extra work occurs at every project. If the extra work is not clearly the fault of the architect or the contractor, the owner pays for it.

Sometimes the contractor or architect will say: "Why worry about it? It has to be done." Or, "You, the owner, want it. So, if the extra work would have been included in the drawings and specifications from the beginning, you would have to pay for it anyway, right?" Not quite!

Almost any work in addition to the contract, after bids have been submitted, costs more than if it had been part of the original contract, simply because the contractor does not bid the extra work competitively anymore. He is already on the job. Who else could do the extra work—a second contractor?

Extra work crops up, inevitably, on every project. Before we tell you what the remedy is, let's look again at the extras and determine why they should be paid by the owner.

a. Unstable soil: Could have been detected only by a soil investigation, with at least two drill holes—too costly for a small project. Better to wait and see if it's required when trenches are dug and then pay the extra cost, a common construction practice. The following notice is even included in our specifications:

"Contractor will bid foundations as shown on drawings. If the Building Official or architect will require a soil report, larger or deeper footings, or any other foundation adjustment during construction, the owner will pay for it."

Needless to say, a soil report should be ordered when a need is clearly established, before design is started.

b. & d. If drawings and specifications are sufficient, but the owner decides to upgrade or increase the work, then it is the owner's obligation to pay. For example upgrading from aluminum to wood windows. The owner says in effect: "I am doing this remodeling project only once. I might as well install the windows I always wanted—wood windows." Architects and contractors refer, with a meek smile, to this type of change as the owners "Might-As-Wells."

c. Corroded plumbing: Hidden conditions can't always be anticipated, but should be allowed for, especially in remodeling work. In case a dispute erupts about the cost of the extra work, the general contractor could tell the plumber, "Just connect it and let the owner find out in a year or two." Obviously, it is wiser to replace corroded pipes while the walls are open and easily accessible. A prudent client does not force the contractor to cover up.

e. Here the architect tried to save the owner money; there were already enough hose bibbs on the house. He had used his best judgment. The owner, not in the planning stage, but during construction, decided otherwise. He should pay for this extra work.

A warning: If contractors or architects are forced by hardnosed owners to pay for mistakes, small or large, for which they are not responsible under common construction practice, they can easily start trading to compensate: "Put in the hose bibb and use a cheaper shower pan." The owner gets his bibb, but how much does he know about shower pans?

How can you not get upset about extras? Anticipate extra work and have a contingency clause in your specifications, which might read like this:

Contingency Allowance:
Base bid to include 8 percent of bid for contingencies. Such extra work has to be authorized by the owner or his agent by a written change order and will be paid from this allowance. Unused portions of money shall be returned to the owner before final payment, either by check or by a deduction from the last payment.

All bids received should include such a contingency clause. For remodeling work the amount should be from 8 to 10 percent.

The $24,500 bid should have had, at say 8 percent, a contingency of $24,500 X 0.8 = $1,960. The bid as submitted should have been: $24,500 + $1,960 = $26,460. $1,725 was spent on extra work and $235 would have been returned to the owner by check or by deducting it from the last payment to the contractor.

DO:

● Anticipate extra work with any project.

● Include a contingency allowance clause in your specifications or agreement with the contractor:
  3 percent for new construction projects
  8 to 10 percent for remodeling

● Authorize extra work only by a written change order. Have the change order signed by owner, contractor and architect.

● Make sure you get back the unused portion of the contingency allowance, either by check before final payment, or as a deduction from the final payment.

● Install as many energy-saving features as possible, more than the codes mandate. Initial cost: somewhat higher than standard construction. Long term: they really pay.

So far, you have decided to build, you know "what it takes" and how a tentative cost guestimate is established. The next question is: should you hire a professional to do the drawings? And whom?

A-6 WHO NEEDS AN ARCHITECT!

2000 years ago, architect Marcus Vitruvius explained his services to the Roman Senate in three words: Commodity, firmness, delight! Not much has changed in 2000 years. Materials, yes, and technology, but not the principles:

Commodity: The building must accommodate the owner's requirements in regard to size, layout, budget and life expectancy.

Firmness: The building must meet the legal and the owner's requirements to guard safety and health.

Delight: The building should be comfortable and have sufficient heating, cooling, lighting and ventilation. It should have a pleasant design and fit into the surroundings.

Do you need an architect for this?

Legally you don't need one for certain buildings: residential remodelings, houses, smaller apartments, simple commercial structures. Regulations vary from region to region, but the buildings just mentioned are exempt in most states from an architect's signature, provided they are: not over two stories, not of masonry and don't have clear spans larger than 25 feet. In most states, even in those cases, the owner could hire an engineer licensed in that state to do the structural aspects, sign the structural calculations and the structural detail drawings. So, what is it the architect can do which the owner cannot?

The situation is similar to the following: Let's say you run a company that grosses $3 million and is at the end of its fiscal year, liable for say $136,000 in taxes—are you able to set up the books correctly, do a reliable bookkeeping job, be aware of all the tax laws, pay as little tax as legally possible and keep the IRS happy (or at least at a distance)? Are you able to do all this besides your regular occupation (assuming you are not an accountant)?

Maybe you can—probably not. A competent accountant will save you more on taxes than he charges for his services; therefore, his services come free. To complete the analogy: What about the $136,000 building project? Can you design it? Be aware of all the building codes, know whom to invite to bid? Evaluate the proposals? Get the lowest possible bid? Avoid lawsuits? Shave construction costs to the bone? Oversee construction? Complete construction on time? Enforce the contract and its guarantees? Get the contractor back after one year to fill the cracks and revamp the faulty heating system? Do you have the time, the knowledge?

Let's consider just **one** of the above facets in which an architect can save money. The one and only way to get the lowest possible responsible bid is through **competitive bidding**. This means bids from more than one contractor—at least three are advisable. Then you must compare the bids. This means, you must get bids based on the same requirements. That is a big task: drawings and specifications must be precise, clear and detailed and yet not discourage the small contractor from bidding.

To produce such drawings and specifications requires an experienced person; nobody can do it if he has not done it many times before. Many people think they can, but they cannot. Even for the smallest project, it pays in most cases to hire someone who knows—an architect, a building designer, a draftsman. If the contractor does the drawings, you cannot get competitive bidding. As a general rule, he will not invite his competitors; he will not release "his" drawings for duplication.

Hire an architect? Remember the comparison with the accountant, who not only keeps you out of trouble, but, in addition, brings your taxes down enough to pay for his services. If an architect can get your building cost down 10 percent, from $136,000 to $122,400, he will save you $13,600! For around 10 percent of $122,400 (or $12,240), the architect does all the work described above. He saves you $13,600 and charges $12,240. You still have $1,360 left over. (Not to mention all the time, energy and worry the architect absorbs for you.)

Your best choice: An experienced architect or building designer who:
► specializes in the type of building you want to erect,
► is eager to take on the project,
► has an office in town,
► offers a reasonable fee arrangement (See A-7),
► steers the project through all government agencies,
► enforces the agreement with the contractor up to the-year-after inspection,
► provides certain legal protection
► saves you about 10 percent of the construction cost.
► In addition to all the above, provides you with a better designed building.

How do you find a good architect or building designer?

● Ask friends or business associates who might have worked with architects or building designers. Ask contractors, realtors, loan officers at the bank, building committee members of churches, schools or clubs you or your friends know.
● Where you see buildings under construction, inquire about the architect. (At the same time, get the names of the general contractor and subcontractors). Read local construction reports ("Daily Construction Reporter") and look for an announced project similar to yours. Normally it lists architects and contractors.
● Phone the local professional chapters of architects and/or building designers.
● Leaf through building magazines and local monthly periodicals.

Good drawings and specifications are necessary! Architects and building designers produce them. Who else? What is the best way to get drawings and specifications? The next chapter will tell you.

A-7 BEST WAY TO GET DRAWINGS

Before you order drawings or start making them up yourself, find out whether you need any at all. If you are building a 2-foot high garden wall, or replacing a window with a sliding door, you probably do not need drawings—you probably don't even have to take out a permit. Check with your Building Department.

If your work does require a **permit,** you may still be able to do without **drawings;** sometimes descriptive notes on the permit form or a city standard drawing attached to the permit will suffice. The Uniform Buildng Code states:

Permit required

No person, firm or corporation shall erect, construct, enlarge, alter, repair, move, remove, convert or demolish any building or structure . . . or cause the same to be done without first obtaining a separate building permit for each such building or structure from the Building Official.

(But there are exceptions, such as the low garden wall.)

Drawings and specifications required:

With each application for a building permit, and when required by the Building Official for enforcement of any provisions of this code, two (or more) sets of drawings and specifications shall be submitted. The Building Official may require drawings and specifications to be prepared and designed by an engineer or architect, licensed by the state to practice as such.

Exceptions: When authorized by the Building Official, drawings and specifications need not be submitted for the following:

1. One-story conventional wood stud construction with an area not exceeding 600 square feet.
2. Private garages, carports, sheds and agricultural buildings used as accessories only and not exceeding 600 square feet.
3. Small and unimportant work.

So, there are situations where you don't need drawings and specifications. But, if you do need them, now is the time to determine the purpose they will serve: To visualize the project in your own mind, for use by the Building Department, the lender, the contractor, or for all these objectives.

Let's take a closer look.

In the **planning and permit phase,** good drawings are needed to show what, how much and at which quality level you want to build. To get a permit, certain standards must be met as spelled out in the Uniform Building Code, Zoning Regulations and other governing laws.

In the **financing, bidding and contract phase:**

a. To find out how much financing you need, how much you can get and at what terms, preliminary drawings will suffice.
b. To obtain the loan, complete drawings and specifications are required. (Lenders have been known to be interested in who prepared the drawings and specifications.)
c. To get reliable and comparable bids, from which you can select the best bidder, watertight drawings and specifications are required and sometimes additional information in the form of addenda.

In **the construction phase,** drawings and specifications undergo a transformation— they become part of the construction **contract.** This is something most individuals do not realize. Yet this very event can become the source of disappointments, misunderstandings and even lawsuits. Think about it: how much of your building project is described in the contract form itself? Hardly any. The price, yes; the required time of completion, maybe. About the building itself, there is very little information. You can conclude that the drawings and specifications are 80 percent of the contract with the contractor. If something is shown on the drawings, you get it, at least you are contractually entitled to it.

If not indicated on the drawings or in specifications, even your attorney cannot get it for you, because it is not a part of your contract!

It is not the appearance, neither the number of sheets, nor myriad notes, numbers or dimensions, but a specific correctness of drawings and specifications (which differs from job to job) that will determine the success of your building project. What, then, is the best way to get drawings? The most economical, most convenient or smartest way?

With no obligation to anyone—contractors, plan mills, package dealers, the A.I.A.— with no favors to anyone and no axes to grind, here are seven ways in order of increasing cost (to you, the client) to get drawings.

Stock plans:

From plan services or building magazines. A best buy IF the drawings are exactly what you want: size, orientation, terrain, level of quality, appearance and cost range. **And** if the Building Department will approve them. Many stock drawings are produced in the eastern United States and do not comply with West Coast building codes. That, of course, could be adjusted. But if there are major changes (i.e., you have to rearrange the living room away from the north side, for instance, or your lot shows considerable slope, but the drawings do not) stock plans are not a good buy, but a headache. If smaller adjustments have to be made, order a "reproducible print" (also called ozalid or sepia). Make changes on the reproducible prints, then have them reproduced by a local blueprinter.

What qualifies as small adjustments? Changing the location of windows or doors, changing the type of exterior siding, adding electrical outlets, upgrading plumbing fixtures, all are easy, small amendments. If structural modifications are necessary, or you want to build one story higher or revamp the roof, forget stock plans. Also, stock plans are of no use for remodeling projects. And don't buy any mirror-reverse prints.

Local stock plans:

An unpublicized source, but they could be a best buy! Visit better residential developments in your area. If you like a particular model under construction or just completed, ask for the name of the designer, very likely a local architect. Make an appointment and try to buy a set of the drawings. He may even make some of those smaller changes for you, mentioned earlier. The question of ethics in "reselling an original design" normally does not come up with development houses or apartments. Most residential development designs for houses, duplexes, townhouses or condominiums are not "original," created solely for one developer. More often than not, the design is adopted from another plan, restyled, revamped, rearranged. It is better to have a good copy than a mediocre original.

Three more hints:

- Stock plans, if they are suitable at all, work best for houses and small apartment buildings.
- If you order stock plans, also order specifications and quantity take-offs. Before you buy, verify that the specifications and drawings are approvable by all government agencies in your town.
- When inquiring about the designer, get also the names of contractors or subcontractors—they may give you bids. Remember, they already know the building!

Hire an architect:

Probably your best buy, even for smaller projects. Yes, residences and additions included. Under Chapter A-6 you read **what he can do for you.** Again, briefly, he should:

- ► Be experienced, even specialized, in the type of project you want to build.
- ► Be enthusiastic and add some unique ideas. Be a good listener. (Watch him listen!)
- ► Offer a reasonable fee arrangement.
- ► Produce the best bid, saving you more than his fee.
- ► Protect you legally, to a degree (compliance with codes, lien releases, contractors' guarantees).

Read A-6 again (WHO NEEDS AN ARCHITECT?) which also tells you how to find one.

For fee arrangement, we recommend:

- ► Schematics and preliminaries done at an hourly rate.
- ► Construction documents (drawings, specifications, addenda and contracts) at a fixed cost.
- ► Construction Phase: "Administration of the contract for construction"—hourly rate or fixed fee.

If you cannot decide on one particular architect, ask two or three to produce some schematic sketches, pay for them, then take your pick. At this stage all preliminary drawings have a value and can be used for input.

Employ a draftsman or a drafting service:

This seldom works, except for the smallest projects—let's say a residence or small apartment building. Even then, if the arrangement is to work at all, the draftsman must be outstanding, not so much as a drafter of lines, but as a technician and negotiator. It is not a matter of personality but one of experience—after all, architects started out as draftsmen. Most draftsmen do not have experience on **all** aspects of a job, although the apparent neatness and quantity of the drawings or their confident personality may conceal these professional shortcomings.

Other drawbacks: Draftsmen cannot be held legally responsible since they are not permitted to sign larger projects. They often are not available during the day for job inspections.

If you do employ a draftsman, choose the best you can find. For a small project, you might find one who works in a responsible position in an architect's office. When that firm temporarily has a low workload, they may not mind him taking on your project, if it's only a short moonlight operation. A **drafting service** is an established business to furnish drawings and specifications. It is normally run by a draftsman; except that he may be available during business hours, the observations made about draftsmen apply. Excellent drafting services are the exceptions.

Contractor provides free drawings:

. . . as advertised. You've seen the ads: Free design, free permits, free financing help. No, drawings will not come free. Most likely the contractor will have the design, and later the permit drawings, prepared by a draftsman, drafting service or architect, freelance or in-house, and you will pay for them as part of the building contract costs—THEY ARE NEVER FREE.

Also, with a contractor who provides the drawings, you will have a negotiated contract since you cannot use "his" drawings to get bids from other contractors. This is the drawback, no competitive bids!

An exception is if the contractor agrees to sell you the drawings and you have the right to duplicate them and to get other bids or even decide to become an owner-builder. In most cases, however, the drawings of package contractors will not be complete enough for competitive bidding since they are not drawn for that purpose. It may sound that we, as architects, knock "design-build packagers." We do not. We are package builders ourselves, for our own accounts. We have designed and built hundreds of thousands of dollars worth of construction. We know the system, including the disadvantages to the lay person. "Design-build" drawings will have minimum information—not enough to get true competitive bids. It is also questionable if enough time is spent on design. This last objection is easily concealed—when the package builder promises to an innocent client to provide design and permit drawings, the client thinks, like Gertrude Stein, "A drawing is a drawing is a drawing—what more do I need?" You need more. Once you spend thousands (or tens of thousands) of hard-earned dollars, strive for a quality project-excellent, unique, with a few special features—not just a flat-roofed box, with some dumb high window, slapped lovelessly against your house's rear end!

Prefabricated building:

No drawings needed—just a screwdriver. A splendid concept, but seldom carried out in practice. Why?

- Same reasons as those against stock plans: size, orientation, terrain, level of quality, appearance.
- High freight costs—most prefabs are trucked in from out-of-town. Your savings could be offset by this added expense. Check!
- Who steers drawings and specifications through the Building Department? Can local workers erect the structure or do you have to hire the out-of-town crew and pay their travel expenses? Who furnishes the foundation, utility connections, sewer lines, final grading?

With a prefabricated building, you only save if it is a small vacation home or if the prefabber does the entire work—foundation, utility connections, grading, obtaining permits—everything.

Draw it yourself?

It might work with simple projects such as decks, carports, garden walls, patios, but—and I know you'll find this hard to believe—it can be the costliest way to get drawings. We have observed that in nine out of ten cases, self-drafted drawings are not sufficient, for various reasons. However, if you have time, you may want to give it a try. Why? If later you decide to get outside help—a draftsman, architect or contractor—the work you have done up to that point helps give you better insight when you work side-by-side with a professional.

Not to scare, but to guide you: obtain a sample copy of a "Residential" or "Commercial" Plan Check Correction Sheet from your Building Department.

On smaller projects, especially with residential or commercial remodeling or additions, the owner or committee members can do certain initial work and thereby reduce professional fees:

► Measure existing structures.
► Sketch them to scale, on graph paper.
► Obtain information on existing utilities.
► Most importantly, write a tentative building program:
  (a) what you think you need,
  (b) future requirements,
  (c) how much you can finance.

Summing up:

BEST WAY TO GET DRAWINGS:

1. Find out whether you need drawings,
2. If so, determine what the drawings will accomplish for you (see previous chapter):
  ► Good design
  ► Desired level of quality
  ► Approvals and permits
  ► Watertight bidding documents
  ► The BEST bidder
  ► Certain legal protection
3. Select the most suitable way to get drawings. Whichever source you choose, get the best drawings you can. So much depends on drawings and specifications:
  Efficient design
  Complying with all code requirements
  Avoiding legal problems
  Bottom line cost of construction
  Lasting satisfaction

In this chapter we talked about drawings and **specifications.** We hope you are curious by now: Who needs specifications?

A-8 WHO NEEDS SPECIFICATIONS?

Love and marriage,
horse and carriage,
drawings and . . . specifications!
The story unfolds:

An elderly neighbor had planned to add a bedroom and bath to her one-story house. A draftsman had made a simple drawing. She took out a permit and received proposals from two remodeling contractors. She liked one of the contractors better since he yielded to her pressure and pleading to get the bid lower and still a bit lower. She was a widow, she told him, and had only a certain amount of money. All she wanted was a cozy bedroom and a nice bath. Nothing fancy, but nice. The contractor seemed to understand.

The project became the disaster of the neighborhood. Tears, hand-wringing and loud, very loud discussions. She called me one Sunday for help. What should she do? What should she doooo?

Let's take an inside look at what happened! The widow had visualized a very nice bathroom, similar to what she had seen in "House Beautiful." Her dream bath was to have wall-to-wall carpet, a wall-hung pullman, cultured marble tops, French doors to the garden, a chandelier on a dimmer switch, the American Standard luxor toilet in a color to match her towels, the shower with marble walls, a captain's brass showerhead, shampoo shelf, a vapor-proof light (on a dimmer switch), a brushed-gold door.

Unfortunately, the contractor could not visualize all the beauty and elegance, because the drawings did not show that. There was no word about the carpet, the marble top, French door, a dimmer switch, marble walls in the shower, the shampoo shelf. Where the shower was to be, for instance, all the contractor saw on the print was a square area labeled "shower"—when he scaled the square, it was 3' x 3'.

Another kink in the project developed which didn't help the situation. When the contractor reached the point of installing the roof, he realized that he had bid the beautiful bath project too low. The interior finish really had to be "skinflinted" if he wanted to come out alive. The shower was the first victim of cost-saving.

He unloaded from his truck one morning the absolute minimum, a flimsy one-piece metal shower—3 feet by 3 feet, as called for on the drawing. At this point the distraught widow phoned us for help.

What was wrong? The drawings? Not really; the prints called for a 3 foot by 3 foot shower and that's what she got, a 3 foot by 3 foot metal shower.

The point is: drawings are not enough. Even if these particular drawings would have been more complete, there is not enough space on any drawing to show more than a square or a rectangle, with some notes such as:

3 foot by 3 foot shower stall
shatter-resistant shower door
furr ceiling to 7 feet

You won't find much more on most floor plans.

Some designers include an interior elevation of the shower front, with additional information: the height of shampoo shelf and soap dish for instance. How do you make sure you get a color-anodized shower door, superior faucets, the adjustable showerhead with water saver, quality ½-inch marble with grime resistant joints?

The answer: Written specifications. They provide space to describe all the items of the shower, the bathroom, the building. Specifications serve several purposes:

- For the owner in the planning stage: As a sort of notebook of what he wants in a building.
- For bidding contractors: To list what must be included in their bids and to prevent that low-skill tasks will be performed by high-priced workmen. (For instance: a laborer, under supervision, should backfill the sewer line trench, not the plumber).

To continue with the shower example, the following information would appear in the specifications:

Under "Lath & Plaster"

Areas to receive simulated marble:	Cement plaster scratch and brown coat over self-furring metal lath on 15 lb. felt.

Simulated Marble

Simulated marble shower walls:	½" simulated marble, color selected by owner from standard samples. Each wall in one piece. Shampoo shelf: ⅝" thick, 12" across, finish top, bottom, exposed edge.
Grout:	Color to match darker parts of marble, with grime resistant adhesive.

Plumbing

Toilet:	A/S Luxor 2003.010 w/custom-styled closed front seat and cover. Color: fawn beige.
Lavatory:	A/S Avalon 0493.015, 22 x 19 w/4" center. Heritage faucet 2103455
Shower floor:	P/F Cascade 36x36, min. flexural strength 4000 PSL. Color: golden brown, slip-resistant surface. Stainless steel drain w/leakproof connection & removable strainer.
Shower fitting:	A/S Aquarian II 1490.051, with Astrojet showerhead 1413.020 w/volume regulator.

Bldg. Specialties

Shower door:	Gold-anodized alum. with towel bar and pivot hinges. Shatter-resistant plastic.

Electrical

    Light fixture in shower: Recessed, stainless steel, 100 W, water and vapor-proof.

    Specifications also include notes on standards of workmanship, installation, coordination with other trades, guarantees and clean-up.

    The shower specifications given as an example illustrate that there is not enough space on drawings to list all the information on every item of the building from foundation reinforcing to roof tiles, from windows to light fixtures, water heaters to doorknobs. There is not enough space on drawings to describe all the materials and to nail down the standards for workmanship, installation, coordination with other trades and guarantees.

    The message:

- You need separate, written specifications.
- For outline specifications of any project or even for final specifications of small jobs, use the 4-page F.H.A.-type and "Description of Material" shown on the following four pages. Although convenient to use, the form does not allow enough space for standards of workmanship, installation, coordination, clean-up, guarantees. Therefore, we recommend it mainly for outline specifications to be used with lenders and for tentative cost estimates by the square footage method.

> Costs of drawings, specifications and construction are not the only ones the owner has to anticipate. There will be incidental expenses, which are outlined in the next chapter.

☐ Proposed Construction

☐ Under Construction

No. ....

Property address .... City .... State ....

Mortgagor or Sponsor ....
                                    (Name)   (Address)

Contractor or Builder ....
                                    (Name)   (Address)

INSTRUCTIONS

1. For additional information on how this form is to be submitted, number of copies, etc., see the instructions applicable to the FHA Application for Mortgage Insurance or VA Request for Determination of Reasonable Value, as the case may be.

2. Describe all materials and equipment to be used, whether or not shown on the drawings, by marking an X in each appropriate check-box and entering the information called for in each space. If space is inadequate, enter "See misc." and describe under item 27 or on an attached sheet.

3. Work not specifically described or shown will not be considered unless required, when the minimum acceptable will be assumed. Work exceeding

minimum requirements cannot be considered unless specifically described.

4. Include no alternates, "or equal" phrases, or contradictory items. (Consideration of a request for acceptance of substitute materials or equipment is not thereby precluded.)

5. Include signatures required at the end of this form.

6. The construction shall be completed in compliance with the related drawings and specifications, as amended during processing. The specifications include this Description of Materials and the applicable Minimum Construction Requirements.

1. EXCAVATION:

Bearing soil, type ....

....

2. FOUNDATIONS:

Footings: Concrete mix .... Reinforcing ....

Foundation wall: Material .... Reinforcing ....

Interior foundation wall: Material .... Party foundation wall ....

Columns: Material and size .... Piers: Material and reinforcing ....

Girders: Material and sizes .... Sills: Material ....

Basement entrance areaway .... Window areaways ....

Waterproofing .... Footing drains ....

Termite protection ....

Basementless space: Ground cover .... Insulation .... Foundation vents ....

Special foundations ....

....

3. CHIMNEYS:

Material .... Prefabricated (make and size) ....

Flue lining: Material .... Heater flue size .... Fireplace flue size ....

Vents (material and size): Gas or oil heater .... Water heater ....

4. FIREPLACES:

Type: ☐ Solid fuel; ☐ gas-burning; ☐ circulator (make and size) .... Ash dump and clean-out ....

Fireplace: Facing ....; lining ....; hearth ....; mantel ....

5. EXTERIOR WALLS:

Wood frame: Grade and species .... ☐ Corner bracing. Building paper or felt ....

  Sheathing ....; thickness ....; width ....; ☐ solid; ☐ spaced ...." o. c.; ☐ diagonal; ....

  Siding ....; grade ....; type ....; size ....; exposure ...."; fastening ....

  Shingles ....; grade ....; type ....; size ....; exposure ...."; fastening ....

  Stucco ....; thickness ....". Lath ....; weight .... lb.

  Masonry veneer .... Sills .... Lintels ....

Masonry: Facing ....; backup .... thickness ....". Bonding ....

  Door sills .... Window sills .... Lintels ....

  Interior surfaces: Dampproofing, ... coats of ....; furring ....

Exterior painting: Material ....; number of coats ....

Gable wall construction: ☐ Same as main walls; ☐ other ....

6. FLOOR FRAMING:

Joists: Wood, grade and species ....; other ....; bridging ....; anchors ....

Concrete slab: ☐ Basement floor; ☐ first floor; ☐ ground supported; ☐ self-supporting; mix ....; thickness ....";

  reinforcing ....; insulation ....; membrane ....

Fill under slab: Material ....; thickness ....".

7. SUBFLOORING: (Describe underflooring for special floors under item 21.)

Material: Grade and species ....; size ....; type ....

Laid: ☐ First floor; ☐ second floor; ☐ attic .... sq. ft.; ☐ diagonal; ☐ right angles.

8. FINISH FLOORING: (Wood only. Describe other finish flooring under item 21.)

LOCATION	ROOMS	GRADE	SPECIES	THICKNESS	WIDTH	BLDG. PAPER	FINISH
First floor							
Second floor							
Attic floor......... .... sq. ft.							

26. INSULATION:

LOCATION	THICKNESS	MATERIAL, TYPE, AND METHOD OF INSTALLATION	VAPOR BARRIER
Roof			
Ceiling			
Wall			
Floor			

27. MISCELLANEOUS:

(Describe any main dwelling materials, equipment, or construction items not shown elsewhere) :

HARDWARE: *(Make, material, and finish)*

SPECIAL EQUIPMENT: *(State material or make and model.)*

Venetian blinds .... Number .... Automatic washer ....

Kitchen range .... Clothes drier ....

Refrigerator .... Other ....

Dishwasher ....

Garbage disposal unit ....

PORCHES:

TERRACES:

GARAGES:

WALKS AND DRIVEWAYS:

Driveway: Width .... Base material ....; thickness ......". Surfacing material ....; thickness ......"

Front walk: Width .... Material ....; thickness ......". Service walk: Width ...... Material ....; thickness ......"

Steps: Material ....; treads ...."; risers ....". Cheek walls ....

OTHER ONSITE IMPROVEMENTS:

(Specify all exterior onsite improvements not described elsewhere, including items such as unusual grading, drainage structures, retaining walls, fence, railings, and accessory structures.)

LANDSCAPING, PLANTING, AND FINISH GRADING:

Topsoil ...." thick: ☐ Front yard; ☐ side yards; ☐ rear yard to .... feet behind main building.

Lawns *(seeded, sodded, or sprigged)*: ☐ Front yard ....; ☐ side yards ....; ☐ rear yard ....

Planting: ☐ As specified and shown on drawings; ☐ as follows:

.... Shade trees, deciduous, ..." caliper.   .... Evergreen trees, ...' to ......', B & B.

.... Low flowering trees, deciduous, ...' to ......'   .... Evergreen shrubs, ...' to ......', B & B.

.... High-growing shrubs, deciduous, ...' to ......'   .... Vines, 2-year ....

.... Medium-growing shrubs, deciduous, ...' to ......'

.... Low-growing shrubs, deciduous, ...' to ......'

IDENTIFICATION.—This exhibit shall be identified by the signature of the builder, or sponsor, and/or the proposed mortgagor if the latter is known at the time of application.

Date .... Signature ....

Signature ....

21. SPECIAL FLOORS AND WAINSCOT:

	LOCATION	MATERIAL, COLOR, BORDER, SIZES, GAGE, ETC.	THRESHOLD	BASE	UNDERFLOOR
FLOORS	Kitchen....				
	Bath....				

	LOCATION	MATERIAL, COLOR, BORDER, CAP. SIZES, GAGE, ETC.	HEIGHT	HEIGHT AT TUB	HEIGHT AT SHOWER
WAINSCOT	Bath....				

Bathroom accessories: ☐ Recessed; material ....; number ......; ☐ attached; material ....; number ....

22. PLUMBING:

FIXTURE	NUMBER	LOCATION	MAKE	MFR'S FIXTURE IDENTIFICATION NO.	SIZE	COLOR
Sink....						
Lavatory....						
Water closet....						
Bathtub....						
Shower over tub*....						
Stall shower**....						
Laundry trays....						

*☐ Curtain rod   **☐ Door   ☐ Curtain rod

Water supply: ☐ Public; ☐ community system; ☐ individual (private) system. ★

Sewage disposal: ☐ Public; ☐ community system; ☐ individual (private) system. ★

★Show and describe individual system in complete detail in separate drawings and specifications according to requirements.

House drain (inside): ☐ Cast iron; ☐ tile; ☐ other .... House sewer (outside): ☐ Cast iron; ☐ tile; ☐ other ....

Water piping: ☐ Galvanized steel; ☐ copper tubing; ☐ other .... Sill cocks, number ....

Domestic water heater: Type ....; make and model ....

recovery .... gph. 100° rise. Storage tank: Material ....; capacity .... gallons.

Gas service: ☐ Utility company; ☐ liq. pet. gas; ☐ other .... Gas piping: ☐ Cooking; ☐ house heating.

Footing drains connected to: ☐ Storm sewer; ☐ sanitary sewer; ☐ dry well. Sump pump

23. HEATING:

☐ Hot water. ☐ Steam. ☐ Vapor. ☐ One-pipe system. ☐ Two-pipe system.

☐ Radiators. ☐ Convectors. ☐ Baseboard radiation. Make and model ....

Radiant panel: ☐ Floor; ☐ wall; ☐ ceiling. Panel coil: Material ....

☐ Circulator. ☐ Return pump. Make and model ....; capacity .... gpm.

Boiler: Make and model .... Output .... Btuh.; net rating .... Btuh.

Warm air: ☐ Gravity. ☐ Forced. Type of system ....

Duct material: Supply ....; return .... Insulation ...., thickness .... ☐ Outside air intake.

Furnace: Make and model .... Input .... Btuh.; output .... Btuh.

☐ Space heater; ☐ floor furnace; ☐ wall heater. Input .... Btuh.; output .... Btuh.; number units ....

Make, model ....

Controls: Make and types ....

Fuel: ☐ Coal; ☐ oil; ☐ gas; ☐ liq. pet. gas; ☐ electric; ☐ other ....; storage capacity ....

Firing equipment furnished separately: ☐ Gas burner, conversion type. Stoker: ☐ Hopper feed; ☐ bin feed.

Oil burner: ☐ Pressure atomizing; ☐ vaporizing

Make and model .... Control ....

Electric heating system: Type .... Input .... watts; @ .... volts; output .... Btuh.

Ventilating equipment: Attic fan, make and model ....; capacity .... cfm

Kitchen exhaust fan, make and model ....

Other heating, ventilating, or cooling equipment ....

24. ELECTRIC WIRING:

Service: ☐ Overhead; ☐ underground. Panel: ☐ Fuse box; ☐ circuit-breaker .... Number circuits ....

Wiring: ☐ Conduit; ☐ armored cable; ☐ nonmetallic cable; ☐ knob and tube; ☐ other ....

Special outlets: ☐ Range; ☐ water heater; ☐ other ....

☐ Doorbell. ☐ Chimes. Push-button locations ....

25. LIGHTING FIXTURES:

Total number of fixtures .... Total allowance for fixtures, typical installation, $................

Nontypical installation ....

3   DESCRIPTION OF MATERIALS

9. PARTITION FRAMING:
Studs: Wood, grade and species .... Size and spacing .... Other ....

10. CEILING FRAMING:
Joists: Wood, grade and species .... Other .... Bridging ....

11. ROOF FRAMING:
Rafters: Wood, grade and species .... Roof trusses (see detail): Grade and species ....

12. ROOFING:
Sheathing: Grade and species ....; size ....; type ....; ☐ solid; ☐ spaced ...." o.c.
Roofing ....; grade ....; weight or thickness ....; size ....; fastening ....
Stain or paint .... Underlay ....
Built-up roofing ....; number of plies ....; surfacing material ....
Flashing: Material ....; gage or weight ....; ☐ gravel stops; ☐ snow guards

13. GUTTERS AND DOWNSPOUTS:
Gutters: Material........................; gage or weight ....; size ....; shape ....
Downspouts: Material ....; gage or weight ....; size ....; shape ....; number ....
Downspouts connected to: ☐ Storm sewer; ☐ sanitary sewer; ☐ dry-well. ☐ Splash blocks: Material and size ....

14. LATH AND PLASTER:
Lath ☐ walls, ☐ ceilings: Material ....; weight or thickness ....; Plaster: Coats ......; finish ....
Dry-wall ☐ walls, ☐ ceilings: Material ....; thickness ....; finish ....; joint treatment ....

15. DECORATING: *(Paint, wallpaper, etc.)*

ROOMS	WALL FINISH MATERIAL AND APPLICATION	CEILING FINISH MATERIAL AND APPLICATION
Kitchen		
Bath		

16. INTERIOR DOORS AND TRIM:
Doors: Type ....; material ....; thickness ....
Door trim: Type ....; material ....; Base: Type ....; material ....; size ....
Finish: Doors ....; trim ....
Other trim *(item, type and location)* ....

17. WINDOWS:
Windows: Type ....; make ....; material ....; sash thickness ....
Glass: Grade ....; ☐ sash weights; ☐ balances, type ....; head flashing ....
Trim: Type ....; material ....; Paint ....; number coats ....
Weatherstripping: Type ....; material ....; Storm sash, number ....
Screens: ☐ Full; ☐ half; type ....; number ....; screen cloth material ....
Basement windows: Type ....; material ....; ☐ screens, number ....; ☐ Storm sash, number ....
Special windows ....

18. ENTRANCES AND EXTERIOR DETAIL:
Main entrance door: Material ....; width ....; thickness ......". Frame: Material ....; thickness ......"
Other entrance doors: Material ....; width ....; thickness ......". Frame: Material ....; thickness ......"
Head flashing .... Weatherstripping: Type ....; saddles ....
Screen doors: Thickness ......"; number ....; screen cloth material .... Storm doors: Thickness ......"; number ....
Combination storm and screen doors: Thickness ......"; number ......; screen cloth material ....
Shutters: ☐ Hinged; ☐ fixed. Railings .... Louvers ....
Exterior millwork: Grade and species .... Paint ....; number coats ....

19. CABINETS AND INTERIOR DETAIL:
Kitchen cabinets, wall units: Material ....; lineal feet of shelves ....; shelf width ....
   Base units: Material ....; counter top ....; edging ....
   Back and end splash .... Finish of cabinets ....; number coats ....
Medicine cabinets: Make ....; model ....
Other cabinets and built-in furniture ....

20. STAIRS:

STAIR	TREADS		RISERS		STRINGS		HANDRAIL		BALUSTERS	
	Material	Thickness	Material	Thickness	Material	Size	Material	Size	Material	Size
Basement										
Main										
Attic										
Disappearing: Make and model number										

A-9 INCIDENTAL BUILDING-RELATED EXPENSES
some nominal — some not so funny

Book a cruise to the Caribbean Islands: After you paid your travel agent for the adventurous cabin on C-deck, there will be more expenses. Certain "extras," "miscellaneous costs," "services not included," a few "incidentals." Whatever they are labeled, expenses they are. Not that they should spoil your fun. But, really, did you expect so many incidentals? For instance:

- update passport
- vaccinations
- motel Fritzi, your dachshund, in a kennel for 10 days
- pay neighbor's boy to take in your mail
- taxi to the ship, plus tip
- tips on board
- drinks are extra
- so are deckchairs
- land excursions to the natives
- tourist knick-knacks
- handwoven baskets
- also: that worn suitcase won't do—not where you are going; you need new shoes, a new swimsuit. Your mother would agree.

If no one has told you about the "incidentals" you incur when building, it's time you were informed. They are necessary, most of them are even useful, but have you allowed for any in your budget? Let's call them building-related expenses. They're not for the land nor for the structure, yet to put the structure on the land, these extra expenses must be paid. Following is a list of several typical building-related costs. Some are nominal, some are not so funny!

Topographic survey:
Recommended for almost any project. Ask your architect to write a brief description of what the survey should include and what is not needed. Then phone two or three licensed land surveyors for a quotation. Use the principle of competitive bidding to arrive at the lowest possible cost for a professional

survey. Before you order a survey, check with the seller of the property or his broker, there may already be one in somebody's files. Preferred scale of survey for residential projects: ⅛"-1'0 to accommodate the architect's preliminary design. Avoid an engineer's scale such as 1 inch equals 1 foot.

Soils report:

This may be required by the Building Inspection Department or the architect. Again, find out from the seller or broker whether there is already a soils report. For certain properties the Inspection Department will insist that you submit one with your building permit application. For most structures, a soils report is not required, but the architect may advise you that he could save you money on foundations if he had a soils investigation.

To get a good soils report quickly and at the lowest cost, get quotations from soil engineers who have previously worked in the area. They can refer to their files on surrounding properties. If soil is a problem, but not an extreme one, we sometimes ask a soils engineer familiar with the area for a "letter of opinion" and tentative foundation recommendations. Later, during excavation, the engineer will do the actual investigation and update his recommendations. **This is a real cost-saving procedure.** Your architect knows how to handle this.

Architects, engineers consultants:

Although in the end professional services should reduce the cost of the project, the professional fees are partly paid up front, i.e., in the planning stage. Fees vary and should be checked and compared with quotations from several professionals.

Lender's financing cost:

An area that gives you a chance to save a bundle. When inquiring about construction financing, request from the lender a written estimate of all fees and costs: loan fees, escrow charges, credit and title reports, preparation of documents, tax service, interest during construction. Get the booklet, Homebuyers' Guide to Settlement Costs, available free from most lending institutions. Review every item on the lender's estimate and ask for an explanation of any charge you don't understand or do not agree with.

Coast Regional Commission:

This panel has jurisdiction in California over projects near the coastline, in general, 1000 yards from the high water line. It varies according to geographical location. Fees range from $50 for small administrative permits to several thousand dollars for major construction. Phone the local office and inquire. Submit preliminary drawings for an informal check and discussion with the Commission's staff members before proceeding with the project.

Architectural Review Boards:

They charge a plan check fee and a permit fee. Sometimes a bond or retainer must be posted to insure the owner's compliance with all covenants, conditions and restrictions (CC&R's). (See Chapter A-11)

Building Permits:

Normally there are two payments: Plan check fee and permit fee. In certain municipalities, additional permits not part of the package must be paid separately. These include permits for electrical work, plumbing, heating, venting, grading and demolition. Procedures vary from city to city. Find out. When you pick up and pay the permit, you may encounter additional charges: water availability charge (some are outrageous!), the water meter itself, sewer fee, sewer connection inspection fee, park fee (typically $200 per dwelling unit), school board fee, engineering field inspection fees.

Special Reviews and Special Permits:

There are many kinds: Classification of Use, Condominium Conversion, Demolition, Encroachment, Environmental Impact Report (EIR), Federal Aviation Agency, Fences and Garden Walls, Floor Area Exception, Harbor District, Height Limit Exception, Highway Department, Hillside Review, House Moving, Land Development, Lease of City Property, Planned District, Planned Residential Development, Public Health Improvement, Rezoning, Site Grading, Street Vacation, Swimming Pool, Tentative and Final Maps, Variance, Work on Public Property. Of course, you never need all of them! But an architect or engineer needs to help you in this jungle. A-10 gives an illustration.

Bonds:

To insure Bids, Performance and Labor and Materials. There are also bonds to protect the owner in regard to completion time, work on city property, leases, roofing, termites and subcontractors. Relax. Most projects don't require either type. (See B-15)

Fund control:

Required by most lenders. Don't just follow the lender's advice as to whom to use for fund control. Shop around! See Chapters B-5 and C-4 for more than you ever wanted to know about controlling funds. The average cost of fund control is $4.50 per $1000 controlled. But you can get it for less, as explained in B-5.

Contractor's escalation clauses:

Generally, contractors proposals are good for 30 days. Most contractors would even accept your contract after two or three months. Inflation in the California building industry hovers around 4 percent to 8 percent per year. That's about ½ percent per month. Theoretically, the contractor should raise his bid by ½ percent when his proposal is not accepted by the owner within a month. There probably is a note in the contractor's proposal to that effect. If not, don't you bring up the subject, even if months have passed. If he does make it a point, you'll know why. But don't simply agree to pay the accelerated cost. Instead, ask the contractor to find an ingenious way to cut some of his costs (his overhead and profit, for instance). He will probably grin and not say anything more about it.

Insurance:

Very important! You must be concerned about insurance of
1. The General Contractor
2. The Contractors under separate contracts. See C-15.
3. Yourself, the owner
4. Bonds. See B-15.
The importance of insurance in construction cannot be overemphasized. Read Chapter C-15 and, man, follow through. Your life savings can depend on it.

Utility bills during construction:

Another breathtaking subject. During construction, should you pay for water

39

the contractor uses in cement mixers or for the telephone calls he makes to Disneyland! (He promised his kids he would take them there next weekend!) Find out in Chapter C-3.

Blueprinting:

Under A.I.A. contracts, the architect pays for two final sets, the owner pays for all other blueprinting and duplicating of specifications, reports, field orders, etc. Small change? A typical 24-inch by 36-inch print costs, at 11 cents a square foot, 66 cents. For a typical residential remodeling, the printing bill might run from $150 to $350, for an office building from $600 to $1500.

Best arrangement: Owner opens an account with a blueprint firm, architect orders prints, signs vouchers, sends monthly statement with signed vouchers to owner; owner pays blueprint company directly.

Owner should show printing cost on Cost Breakdown Sheet to make the expense a part of long-term financing.

Testing:

Testing of soils, materials, welding will be performed by private, independent testing laboratories and their field inspectors. Since testing is done mainly for the benefit of the owner, he hires and pays the testing firms. Testing is normally needed on commercial projects or where difficult soil conditions exist. Amount of work and rates vary; cost can be substantial. Once you know from the building department what has to be tested, obtain competitive proposals from several firms.

Building Department penalties:

Paid to revive an expired permit. This happens, for instance, if the first official inspection did not take place within 120 days after permit issue or if work did not progress for 120 days. Carefully read the expiration dates of all permits.

Re-inspection fee:

Charged when corrections requested by the Building Inspector are not made in an acceptable manner and the inspector must make an additional inspection of the same item. $10 to $15 per re-inspection is average. Rates vary from city to city.

DO:
- Anticipate incidental building-related costs; scrutinize each one.
- Use competitive bidding to hold these expenses to a realistic minimum.
- Channel the expenses into the long-term loan by including the cost on the construction cost breakdown sheet, which you submit with your loan application.
- If the money you have earmarked for permits is in an interest earning account, take out permits at the latest possible date.

A sample form of a construction cost breakdown is shown on the next page.

The importance and cost of special permits
need further explanation. A-10 and A-11
give you the information.

CONSTRUCTION COST BREAKDOWN

Date _____

Owner _____ Job Location _____

Owner's Address _____ Phone _____

Plans and Specifications _____	FORWARDED _____	
Engineering _____	Insulation—Weatherstripping _____	
Soil Compaction Test—Survey _____	Sheet Metal_____	
Building Permit _____	Masonry—Fireplace _____	
Water Meter_____	Masonry—Other _____	
Sewer Connection _____	Electric—Rough_____	
Excavating—Grading _____	Electric Fixtures _____	
Foundation Labor _____	Heating _____	
Forming Labor _____	Plumbing _____	
Equipment Rental _____	Built-in Appliances _____	
Piers_____	Plastering _____	
Concrete—Rough _____	Scaffolding _____	
Concrete—Finish _____	Painting _____	
Reinforcing Steel _____	Wallpaper _____	
Reinforcing Steel Labor _____	Iron Work_____	
Rough Hardware_____	Overhead Door _____	
Lumber—Framing _____	Window Cleaning _____	
Lumber—Finish _____	Clean-Up—General _____	
Labor—Rough Carpenter _____	Landscaping—Drainage _____	
Labor—Common_____	Water, Electricity, Fuel Gas _____	
Labor—Finish Carpenter _____	Phone _____	
Paneling _____	Compensation—Liability _____	
Frames for Doors _____	Interest _____	
Frames—Misc_____	Bond _____	
Doors _____	Recording—Title _____	
Windows and Screens _____	Contingency Allowance _____	
Glass and Mirrors _____	Fencing _____	
Hardware—Finish _____	Finish grading_____	
Building Specialities _____	Tub enclosure, shower door_____	
Cabinets—C.C. Doors _____	Equipment Rental _____	
Tile _____		
Cabinet Tops _____	Coastal Commission _____	
Linoleum _____	Architectural Review Board _____	
Asphalt Tile—Etc. _____	Special Permits _____	
Shades—Blinds, Etc. _____	Fund Control Fee _____	
Flooring—Hardwood _____	Blueprinting_____	
Roofing_____	Overhead _____	
Water Proofing _____	Profit _____	

FORWARD _____   TOTAL _____

Courtesy of

DIXIELINE LUMBER CO.

3250 SPORTS ARENA BLVD.—P.O. BOX 80307
SAN DIEGO, CALIFORNIA 92110
224-4161

561 NORTH TULIP
ESCONDIDO, CALIFORNIA 92025 - 745-7271
1400 W. 28TH STREET
NATIONAL CITY, CALIFORNIA - 474-4671
1262 EAST MAIN STREET
EL CAJON, CALIFORNIA 92021 - 442-8415

UNIFORM BUILDING CODE

BUILDING INSPECTION DEPARTMENT

TYPICAL FRAME SECTION - SINGLE STORY

MIN. RIDGE 1" x 6" AND NOT
LESS IN DEPTH THAN CUT END
OF RAFTER (for min. rafter
slope of 3 in 12)

CUT RAFTER FOR
FULL BEARING

E TABLE: 25-T-R-1 thru 25-T-R-8

Form Approved. O.M.B. No. 004-R0001

DO NOT REMOVE CARBONS

DEPARTMENT OF TRANSPORTATION
FEDERAL AVIATION ADMINISTRATION

OF PROPOSED CONSTRUCTION OR ALTERATION
(complete both A and B below)

FOR FAA USE ONLY
AERONAUTICAL STUDY NO.

FAA will either return this form or
issue a separate acknowledgement.
proposed structure:

O.C.:
AR ING
USED

A
1

Department of PLANNING AND LAND USE
Codes Division

THIS APPLICATION BECOMES A VALID PERMIT WHEN SIGNED BY AN AUTHORIZED REPRESENTATIVE OF
THE BUILDING INSPECTION DEPARTMENT AND RECEIPT OF PAYMENT OF TOTAL FEES DUE IS
ACKNOWLEDGED BY THE CITY TREASURER IN THE SPACE PROVIDED.

APPLICANT: USE BALL-POINT PEN & PRESS FIRMLY
FILL IN BLANKS WITHIN HEAVY BOUNDARY LINES

SEPARATE ELECTRICAL, PLUMBING AND MECHANICAL PERMITS MUST BE OBTAINED BEFORE
ANY OF THESE TYPES OF WORK MAY BE PERFORMED ON THIS PROJECT.

PROJECT ADDRESS

BUILDING PERMIT APPLICATION

DIST. NO. | CENSUS TRACT NO. | NIT BER

FIRE PREVENTION BUREAU

HAZARDOUS MATERIALS OR PROCESSES APPLICATION

Application is hereby made by the undersigned for a permit to

☐ Use ☐ Store ☐

BUILDING INSPECTION DEPARTMENT

DOCUMENTATION OF UNREASONABLE HARDSHIP

No.

Type of Variance
(zone, yard, setback, fence, etc.)

Application No.

Case No.

Date Page

APPLICATION FOR VARIANCE
From the Provisions of the Planning and Zoning Regulations of

TO THE ZONING ADMINISTRATOR

CASE NUMBER

BOARD OF APPEALS APPLICATION
DEPARTMENT OF BUILDING INSPECTION

1. The Board of Appeals is legally empowered to (1) investigate and advise on the suitability of alternate materials
and types of construction, (2) provide reasonable interpretations of the building laws where the meaning may be
obscure, and (3) recommend new legislation to the City Council. The Board may recommend approval of minor
deviations of the building laws in certain cases.

2. All appeals to the Board must be within the scope of authority described above. Only those items requested in
writing in this appeal will be considered. Any appeal which has been submitted to the Board for consideration
requires action by the Board. Such appeal may NOT be withdrawn by the applicant.

3. Hearings are open for public attendance. You will be notified of the hearing date.

4. Address all communications to: Board of Appeals, Inspection Department, City Operations Building, 1222 First
Avenue, San Diego, California 92101. **THIS FORM MUST BE SIGNED BY THE OWNER.**

INSTRUCTIONS

RE ONLY - PLEASE PRINT OR TYPE

BLOCK | SUBDIVISION

TELEPHONE

UNIT

JOB STATUS

☐ PROPOSED

A-10 CONDITIONAL USE PERMITS, VARIANCES, ENCROACHMENTS, SPECIAL PERMITS AND APPEALS

A Lutheran Church group commissioned an architect for the design of a new sanctuary which was to replace a small fellowship hall. The architect pointed out to the committee that more than the usual drawings, specifications and building permit were required. The proposed location of the sanctuary and the task of joining the new sanctuary with an existing educational building would require a number of special permits. These would have to be secured before working drawings commenced and before a regular building permit could be obtained.

Sometimes one, or infrequently, several of these special permits are necessary. Let's use the new sanctuary project to illustrate when and why applications must be made for special permits.

a. The church was located in a residential neighborhood, zoned for single-family dwellings. Therefore, many years earlier, when the original fellowship hall was erected, the church members had taken out a **Conditional Use Permit.** This document allows the use of the land for purposes other than those laid down in the zoning code. In this case, an assembly building could be constructed where the zoning code allowed residences only. Now an **Amendment** to the Conditional Use Permit had to be obtained.

Between $500 and $1,500 must be paid with the application for such a permit, and it is normally not refunded if the permit is not granted. The architect recommended that he and the members of the church committee meet informally with the zoning administrator to discuss the chances of obtaining the C.U. Permit. After this meeting, the church members could decide whether they wanted to file the application. Later, there would be a public hearing, where the zoning department would rule on the case.

Once granted, the Conditional Use Permit will define in detail the conditions and restrictions under which the property might be used for the intended "special use." Any individual or group must be prepared to meet opposition from neighbors or local groups at the public hearing.

Obtaining the permit consumes time and money: for the application, for the extra work the architect has to perform and for color renderings and slides. But it is easier to obtain this special use permit than to rezone, an alternate method to change the usage of the land.

b. A portion of the existing fellowship hall—soon to be replaced by the new sanctuary—had been erected within the 15-foot front yard. The architect decided he needed that space again for the new building; therefore, he asked the Zoning Administrator for permission to build again within the front yard. A **Variance** permit was required. An application had to be filed and an additional fee paid.

   And there is still more to come!

c. The original fellowship hall not only intruded into the minimum required front yard, but also was erected smack along one alley side. In order to provide a six-inch roof overhang, the new roof would have to encroach six inches onto public property, the alley. You guessed it: The church needed an **Encroachment Permit.** Another application, another fee, another hearing, another permit.

d. To provide access to the enlarged parking lot, a new concrete driveway had to be installed and a portion of the street curb removed. That constituted "work on public property." You're tuned in if you guessed: another application, another fee, another permit. The process may seem overwhelming but credit must be given to city hall where credit is due. All four applications were tentatively reviewed at the same time, filed and heard at the same public meeting.

e. One other problem was a two-hour rated firewall (as required by the Uniform Building Code) where the new sanctuary would adjoin the existing educational building. In lieu of the firewall, the architect proposed an **equivalent** installation consisting of a row of ceiling sprinklers and additional sprinklers in the rooms on either side of the proposed row of sprinklers. The Building Department has no authority to waive requirements of the building code. Therefore, the equivalent solution was presented to the **BOARD OF APPEALS.** The board consists of appointed experts who represent the Building Department, the Fire Marshal's Office and the general public.

   This board has jurisdiction to permit materials and installation methods that it considers equal to those prescribed by the building code. The burden to prove that an alternate method is equivalent is with the owner, i.e., his architect. The board, after discussion, approves or disapproves the proposed solution, sometimes with modifications. If the Fire Marshal's Office **and** the Building Department are in favor of a proposed equivalent, the board members who represent the public will often go along, and in most cases the matter is settled. Not always, however. Trying to get approval without support of the Fire Marshal and the Building Department is almost always futile.

Before filing applications and paying fees for these permits, arrange for informal meetings with the department's personnel to get an indication of the chances for obtaining the permits. Have your case well-documented on the prescribed forms. Also have precise computations, colored prints and renderings.

As mentioned in Chapter A-9, there are more special permits than illustrated above. Your architect will advise you.

To speak effectively at a **public hearing,** the general rules are:
• State your name, address and location of the property.
• State your position briefly and concisely: for or against.
• If you make a request: State your request and support it by precise statements, drawings, diagrams, slides and renderings. Presentation material should be first class.
• State facts within your knowledge and experience. Refrain from gossip, hearsay, repetition, too many statistics.
• Discuss only what relates to the hearing and to the decision you wish the administrator, board or council to make.

- Organize your presentation within the prescribed time limit. Upon completion, refrain from additional comments unless requested by the board. Remember the credo of the Toastmasters Clubs: "Get up—speak up—shut up!"

DO:
- Budget additional time and money to obtain special permits if they are required.
- Prepare for preliminary meetings and public hearings carefully, like an attorney would prepare for a case in court. The approval of your project can depend on its presentation.

CITY OF SAN DIEGO PLANNING DEPARTMENT
CITY ADMINISTRATION BUILDING
202 C Street
San Diego, CA 92101

APPLICATION
CHECK ALL PERMITS YOU ARE APPLYING FOR

☐ CLASSIFICATION OF USE	12☐ MANUFACTURING - INDUSTRIAL PERMIT*
☐ COMPREHENSIVE SIGN PLAN	13☐ PLANNED COMMERCIAL DEVELOPMENT*
☐ CONDITIONAL USE PERMIT	14☐ PLANNED DISTRICT
☐ CONDOMINIUM CONVERSION PERMIT	15☐ PLANNED INDUSTRIAL DEVELOPMENT*
☐ DEMOLITION PERMIT	16☐ PLANNED RESIDENTIAL DEVELOPMENT*
☐ ENCROACHMENT PERMIT	17☐ REZONING*_____ TO _____
☐ FLOOR AREA RATIO EXCEPTION PERMIT	18☐ TENTATIVE SUBDIVISION OR PARCEL MAP
☐ HEIGHT LIMIT EXCEPTION	19☐ STREET VACATION
☐ HILLSIDE REVIEW PERMIT	20☐ VARIANCE
☐ LAND DEVELOPMENT PERMIT	21☐ OTHER (SPECIFY) _____
☐ LEASE OF CITY PROPERTY	

> *APPLICANT MUST FURNISH AN ASSESSOR PAGE OBTAINABLE FROM COUNTY ASSESSOR'S OFFICE OR A PLAT PREPARED BY A LICENSED ENGINEER OR LAND SURVEYOR SHOWING EXACT BOUNDARIES OF THE SUBJECT PROPERTY.

2☐ ENVIRONMENTAL ANALYSIS _____

S ENVIRONMENTAL ANALYSIS
EN DONE ON THE SITE BEFORE   ☐ NO   ☐ YES

QD NO. (if yes) _____

APPLICANT PLEASE NOTE:
- INFORMATION MAY BE OBTAINED BY CALLING THE CITY PLANNING DEPARTMENT AT 236-6460
- TYPE OR PRINT EXCEPT SIGNATURES
- DEPOSITS AND FEES MUST ACCOMPANY APPLICATION
- PLEASE PROVIDE STREET ADDRESS FOR CERTIFIED MAILING P.O. BOX CANNOT BE USED FOR THAT PURPOSE

THIS SPACE TO BE USED BY PLANNING DEPT.

DATE RECEIVED _____

E.Q.D. NO. _____

LAMBERT COORDINATES _____

COUNCIL DISTRICT _____

CENSUS TRACT _____

ZONE DISTRICT _____

SCHOOL DISTRICT _____

COMM. PLAN _____

EXISTING ZONE(S) _____

Item #   Hearing Date   J.O. #

Typical application for special reviews and permits (Shown: Upper part of first page)

For certain projects, another "special permit" is needed: approval by an Architectural Control Board.

Ahhh, the benevolent dictators . . .

A-11 ARCHITECTURAL CONTROL BOARDS

In order to regulate building activity, more and more communities add another layer of government: Architectural control boards. Whatever their name—Environmental Review Committee, Home Owners' Association, Woodlands Protective Organization—they all have the same function, namely to tell you what and where you can and cannot build. In the protracted process of "getting something built" these control boards are one more stop where one must file an application, submit drawings, pay a fee, wait for approval, comply with rules and please inspectors. Heaven, when will it end?

Is it worth the trouble to buy a lot, or buy a house with the thought of remodeling it, in an area where such a board governs with a set of solemn and forbidding CC&R's (Covenants, Conditions and Restrictions)? Is it worth the extra work and sometimes aggravation, to comply with the board's rules, suggestions, whims or downright dictatorial attitudes?

In most cases, the overwhelming conclusion is yes, it is worth it. You see, the average Zoning, Engineering or Building Inspection Department in this country is concerned only with aspects of health, safety and energy regulations—very seldom with the quality of the environment. That's not to say that some officials don't realize that quality, or the lack of it, does affect the health of people; it simply is not in their jurisdiction. The official explanation is, of course, that the government, local, state or federal, cannot regulate aesthetics, except where the public would mandate such an influence. So, where the government agencies drop the ball, the architectural control boards presumedly pick up. They are interested in the appearance, the tasteful atmosphere of the neighborhood.

Some boards are more successful than others; some spend too much time enforcing misguided architectural concepts or enjoin in amateurish knitpicking of submitted plans. There are, however, very effective boards who have created and maintained outstanding communities. The main source of their effectiveness is the high level of public involvement and creativity prevalent in their communities.

Most boards follow a set of written guidelines, the Covenants, Conditions and Restrictions. Unfortunately, many of these are verbose, in cumbersome lawyer's language. CC&R's depend on interpretation by the board; deviations and exceptions are sometimes permitted. However, like all building codes, CC&R's are restrictive. Codes do not tell you what you can do; they give you limitations, spell out what you cannot do.

CC&R's are part of a legally binding contract between the buyer of the property (you) and the community, which is represented by the board. The contract is enforceable in court, but it is not enforced by the city or county inspection department. If, for instance, the county requires a 50-foot front yard, as far as the county is concerned, this defines the line on which your house can stand. If the CC&R's require 85 feet, this really defines the building line for your house; that is, if you want to avoid problems with the architectural board.

If the contractor makes a mistake and allows only 80 feet, the board's inspector might raise hell, but the county building inspector couldn't care less. The county or city is not a party to your private contract.

Not only front yards are regulated by those numerous pages of covenants, conditions and restrictions, but a myriad of other items as well: side yards, rear yards, distances to stables, number of stories of buildings, number of horses per acre, garage doors facing the street, antennas, visible clotheslines, chain link fences, garbage can exposure, noise from motors, barking dogs, Mexican parrots—you name it. CC&R's have page after page of well-intended no-noes. On the last page you are usually informed that you pay all reasonable attorney fees, court costs and related expenses if you decide to be a maverick and ignore the rules.

The authoritarian ring is an outstanding characteristic of the voice—and letters!—of control boards. And yet, we have found most board members to be friendly and understanding folks with above average intelligence and a willingness to compromise here and there. The most successful boards set very high standards and then hire an outside architect to interpret the standards and to recommend changes or trade-offs in individual cases. These boards know, and act accordingly, that rules are only a means; the end is to maintain a high quality environment that everyone can enjoy.

Suggestions for dealing with architectural control boards:

► Before buying property, read a copy of the CC&R's. If you don't like the rules, don't buy property in the controlled area.

► Once you buy the property and decide to build or to remodel, you must make certain submissions to the control board. The CC&R's spell these out. Generally:
   a.   An application.
   b.   Site plan, with landscaping layout, preferably in color.
   c.   Floor plans and exterior elevations, with material description and color schedule.
   d.   A rendering, if you have one available.
   You also must pay a fee for plan review and permit once the drawings are approved. Sometimes you have to deposit a retainer or bond to insure that you will install landscaping within a certain period of time after the building is completed (six months, perhaps). Normally the board also wants a copy of the drawings as submitted and approved by the building inspection department. The board may have your project inspected by a special field inspector to insure compliance with the drawings. Probably there will be three inspections: (a) after foundation trenching, (b) during framing and (c) after completion.

► Most boards have a rule that you hire an architect when you intend to build or remodel. Hire one who is familiar with the area and the CC&R's. Having worked in the area should not be the controlling criteria in choosing an architect, but it is likely that among the architects and consultants who have a name in the area, you can find one who meets your requirements.
   Let the architect file drawings and attend the board's review meetings.
   Drawings are best submitted as brownline prints, they look richer and softer than

blueline. Site plan and exterior elevations—and renderings if you provide them—should be colored. Indicate lush landscaping, and we hope you mean it. If the budget aches, it is better to specify plenty of smaller plants and shrubs (1 gallon), than to show 5 and 10 gallon specimens 50 feet apart.

It is smart to make your presentations first class without trying to "snow" the committee. The best committees hire a good architect as a consultant to review submittals and he will probably be no dummy. Be honest **and** impressive.

► The drawings you submit for approval can be called Design Drawings as opposed to the Working Drawings you will submit later to the building department and which will also be used for bidding, contracts and construction.

Before you get to the forums of city hall, changes and deviations from the design drawings will take place. When you finally submit to the board the one set of working drawings as checked and approved by the building department, point out the changes in a cover letter. The control board will understand.

What causes deviations? The owner wants to change windows or the location of the garage, the architect simplifies the roof line, the electric company changes the transformer location, the zoning department objects to the distance of the driveway from the property line. All right, let's change it. If there have to be major adjustments it is best to resubmit adjusted Design Drawings to the board for approval and then proceed with Working Drawings. Remember, control boards want to feel in control. They get upset when the owner pops up with surprises.

► It pays to have a friendly relationship with the board and its hired staff, including the special field inspector, if there is one. (See C-9, "Cherish Your Inspectors—All of Them".) We want to stress here that architectural control boards wield a strong influence. The more exclusive the area, the stricter the rules. In the end you benefit much more than they—the art is to see it that way and be cooperative!

You have returned home from City Hall, public hearings and your smooth presentation to the Woodsy Country Home Owners' Association.
What's next?

A-12 THE "WHAT'S NEXT" LIST

We were hired to plan and put out for bid a residence in San Diego's North County. In the first meeting with the couple we discussed their building's needs, jotted down information about their lot and looked at pictures of the lot. We talked about approximate square foot costs and some of the client's preferences. We mentioned the types of contracts they might consider, agreed on a time table for planning, bidding and construction, promised to visit the lot and, of course, explained our fees and asked for a retainer.

After we had visited the property (and picked up the latest edition of Covenants, Conditions and Restrictions), we had a second meeting in which we presented two tentative site plans with schematic floor plan sketches.

We also had a list of questions and loaded our clients with homework assignments (for instance to review survey bids, study our schematic designs and begin negotiations with lenders). For his and her convenience we listed our questions and homework assignments on a sheet of 8½ x 11 paper, xeroxed it at the end of the meeting and handed copies to each of them and to our draftsman assigned to this project. Until the time we obtained the building permit, we issued such a list at the end of every meeting. We called it, without literary ambitions, the "What's Next List."

Yes, it was one more piece of paper to handle, but everyone benefited—the clients, the architect, his staff. This single sheet was a handy agenda of information received, items to think about, events to come and questions to be prepared for. One typical list looked like this:

Client	Architect
Review: 2 bids for survey (will be ordered by architect, paid by client)	Order soil report.
Outline specifications	Call Health Dept. re: septic tank.
	Expect survey, superimpose buildings on survey.
Consider: Size-orientation Cost ranges	Call Architectural Control Board for submittal dates for preliminaries.
Think about: Future pool location Whirlpool?	Prepare package for lenders and give to client. Include ext. elevations and bird's eye sketch of site arrangement.
Meet with lenders. Take along: Schematic drawings Outline specifications Probable construction cost tabulations Loan application form Your financial statement	Update cost range estimates. Around February 28, call client to arrange for next meeting.

DO:

- If information and questions seem overwhelming, ask your architect to prepare for you an informal "What's Next" list.

Chapter A-8 emphasized how crucial it is to specify all materials, methods of installation and guarantees—and these were all tangibles. Difficult to specify is one particular **intangible**—workmanship. The next chapter explains.

A-13 WORKMANSHIP SHALL BE THE BEST!

It should be. In five out of six cases it won't be. Ours is not the age of proud guildsmen and patient woodcarvers or even sufficient apprenticeship training. No master builders nor artisans are among today's heroes: mass adoration and high pay go to football players, movie stars and, in construction, to real estate merchandisers. Quality is in low demand. Instead, in demand is quantity, fast turnover and impressive curb appeal, all of which promise a more favorable return on the investment.

No more tears. No illusions either. The truth is, solid training and pride in work, which produce good workmanship, are not the rule but the exception. Yet the going rate is $20 to $35 per hour for construction work. What steps, what precautions can the owner take to get good workmanship for the money he pays? How can he get performance around 7 or 8 on a scale from 0 (lousy) to 10 (superb)? How can he pluck acceptable workmanship out of these fellows with their nailing machines, radios and dogs?

Your architect should know how to handle the problem. If you are interested in understanding his work better, or if you depend on your own wits, here are some methods for procuring good workmanship:

1. **Settle for an architectural style that can live with inaccuracies** such as hairline cracks in the stucco, a few checked beams or waves in fascia boards. Use a rustic style: if you don't know the label, call it "California Mediterranean." Textured stucco, deep roof overhangs, moldings or battens where materials change, exterior surfaces not painted but stained—the average contractor can do a satisfactory job with such design features.

2. **Good materials and equipment are prerequisites for good workmanship.** Unseasoned lumber, with a high moisture content, automatically produces lousy drywall finishes, with bulging walls and popping nails. A cheap bathroom fan makes a racket no matter how well trained the electrician is who installs it. A painter as dedicated as Michelangelo cannot paint even your garage ceiling satisfactorily with a low grade paint. If the finished product—deck, residence, new sanctuary—should be rated around 7 or 8 on our scale, materials and equipment must be suitable for that rating.

3. **Hire competent contractors and workmen.** This is half the solution to the problem of workmanship. Before you hire someone, insist on seeing two or three projects he has

completed. Hiring an incompetent contractor is like going to an incompetent dentist–by the time you sit (or lie down) on his chair, the man had better be good. Don't let him learn on your teeth; remember, you have only one set. The same goes for your building project—don't let it become a training ground for amateur carpenters.

4. **Many building sites look like pigpens, or worse. Not yours!** Tell all trades to pick up their debris, boxes, beer cans, orange peels and wire clippings. Withhold payment until that order is followed. For larger projects, you may have to rent a dump bin. You still have to insist: Debris goes into the barrel or bin, not on the floor or into foundation trenches (which haven't been backfilled yet). The point is: a messy building site is not conducive to good workmanship.

5. **Specifications must emphasize quality workmanship.** Experts agree that it is difficult to word a contract to guarantee good workmanship. The timeworn phrase that this or that should be done in a "workmanlike manner" has no substance; it is meaningless. A judge or jury will ask, "Like which workman? Like the guy who installed your neighbor's wavy fascia?"

   To specify workmanship, use phrases that are recognized trade language. You'll find these phrases under the subsection "Installation and Workmanship" in almost every section of professional specifications. For example:

   "Installation and Workmanship" of Finish Carpentry:

   > On exposed surfaces work shall be finished smooth, free from machine and tool marks, abrasion, raised grain. Do not install trim, millwork, cabinetwork until plaster or drywall has thoroughly dried.
   >
   > Running flat finish: With cerfed or hollow back.
   >
   > Joints: Tight and formed to control shrinkage.
   > Where piecing is required: Miter; no butt joints.
   > Corners of molded members: Mitered.

   Instead of listing features or procedures to achieve good workmanship, you can refer to recognized standard specifications that contain subsections on workmanship and installation. For instance:

   > All wood shake work shall conform to the latest specifications of the Red Cedar Shingle Bureau.

6. **We strongly advise against acting as your own contractor,** except for small work. Yet, if you do, insist that all trades check and approve the work performed by the preceding trade. The drywall hanger shall approve of the framing, the ceramic tilesetter shall check the work of the drywaller, the painter shall check the finish carpentry and cabinetwork. If conditions cannot be approved, the contractor or you should call the previous trade back. Don't pay anyone whose work is not completed and accepted.

   Except under unusual deadline pressures, no tradesman shall be allowed to work outside after daylight or inside under makeshift lighting conditions. (We once saw a garage drywalled, taped and spackled in the light of the drywaller's truck headlights. The result was a disaster.)

DO:
- Settle on an architectural style that can live with inaccuracies.
- Specify material and equipment of good quality.
- Hire competent contractors or workmen.
- Keep the building site uncluttered.
- Specify good workmanship. Use recognized trade phrases.

- Think twice before deciding to act as your own General Contractor. Coordinating trades requires long experience.
- Have strong guarantee requirements.

DO NOT:
- Pay any contractor or tradesman before his work is completed and accepted by the succeeding trade.

MORE OF THE SAME
On the subject of **quality,** here is some feedback from a recent Conference on Design and Construction. As reasons for poor quality, these were given:
► Inadequate specifications
► Lack of field experience by the design team
► Lack of construction experience by the builder
► Inadequate skills of construction workers
► Contract was given to the low bidder without further checks
► Low architects or engineers fee.

Almost what we said, isn't it?

Now you're rolling! You have drawings, specifications, permits applied for; the architect works within your budget. Soon contractors will submit firm bids. Firm bids mean: you will know how much this extravaganza is going to cost! If necessary, can you increase the budget? If not, what will you do? Decrease the building? A-14 offers you a clever solution.

If the budget is too small . . .

A-14 MAKE PROVISIONS FOR FUTURE INSTALLATIONS

During the first meeting with a client, especially when designing residences or remodeling, the owner should be encouraged to start the three lists mentioned in Chapter A-3: Items the client
1. **Definitely Wants**
2. **Maybe Wants**
3. **Doesn't Want**

On the first list he should put things such as family room with fireplace, double oven, large wardrobes, zoned heating, french doors. On the second list go amenities on which the client cannot make a firm decision—a trash compactor, a fireplace that recirculates air or a built-in vacuum system. The client cannot make a decision because (a) he is not sure he wants the item at all, or more often, (b) he has doubts that his building budget can include it. On the third list, of course, the client specifies what he does not want or like—a 2-story building, fiberglass bathtubs, false beams, Greek columns.

The lists help bring client and architect to a solution more quickly. The client should also be encouraged to bring magazine clippings he may have collected. Why not! A picture can convey to the architect or contractor the client's wishes or dreams better than 499 words.

Not that the architect cannot come up with his own ideas, but the lists help to clarify the client's thinking and that helps him when we start asking questions. We still might suggest that a 2-story house is a good solution in a particular case or that a fiberglass full-height shower, in addition to a bathtub, might be welcomed by someone in the family.

For now, let's take a closer look at the second list, the tabulation of maybes. Besides shelving decisions on these items for a while, the list serves another practical purpose during the planning and bidding stages. When the contractor's bids come in and are finally spread out on the living room table, the best bid might be higher than the client's budget. What do you do now?

Some of the client's wishes and dreams apparently cannot materialize, and it is likely that some of the items that have to be scratched for good are on the maybe list. The trash compactor, for instance, or a built-in vacuum system, or low-voltage lighting, the writing desk in the kitchen or the attic fan. So, when the bids are in, the client looks over the lists again and says, "Sorry, trash compactor, goodbye forever—we don't have you in our budget!" Not quite! Let's not be so hasty.

Here lies the usefulness of the maybe list. If you don't have the funds now, but you really want the trash compactor, for example, **make provisions for future installation.** How is this done? When ordering the kitchen cabinets, tell the cabinet maker to provide a 16-inch wide section with no drawer in the base cabinet. Just a section with a hinged door and two removable shelves is all. The electrician should provide a 110-volt outlet in the rear of that space, 12 inches above the floor. One or two years from now you can buy the compactor, remove the door, plug the cord into the outlet, slide the compactor into the provided space and fasten four screws. Bingo!

It doesn't cost much to make such provisions. In the case of the compactor, the cost of the drawer and drawer hardware is saved and could buy instead the electrical outlet.

The message: When your building budget is strained, look at the maybe list. Make provisions for items you really would like to have in the future but cannot have today. Following are typical items some of our clients could not afford the first time around, but made provisions for installations at a later date:

Deferred Item	Provisions Made
Additional fireplace	If future fireplace will be masonry, install foundation; frame ceiling and roof openings; install cleanout; install gasline and cap it.
Separate shower	Provide a space; install rough plumbing in walls and a drain in the floor.
Air-conditioning	Attached to furnace. Condenser pad; 220-volt outlet near outdoor pad; wire lead to thermostats; condensate line, larger supply and return air ducting; base equipment designed for air-conditioning.
Garage door opener	Intall "operator" door hardware, electrical outlet at ceiling, wire and button.
Intercom or alarm system	Wires in walls or ceilings, with wall terminals (jacks) in designated walls.
Built-in vacuum system	Outlets in walls; tubes or hoses to future motor location; 110-volt outlet at motor location.
Sauna, whirlpool	Rough plumbing and electrical; gasline, capped; vents through roof. Maybe: sleeves through the roof for a future solar panel installation.
Lawn sprinklers	Pipes under concrete walks and driveways; ¾ tee water pipe connection point, if possible ahead of pressure regulator.
Additional rooms	To be finished in future. Leave interiors unfinished, except install insulation and, if possible, electrical outlets and switches; or just have enough spare circuits on a nearby electrical panel. Have points of connections for heating; install windows and doors.

These examples are from new or remodelled residential design and construction work. But the principle of making provisions for future installations applies to many types of buildings: apartments, schools and especially religious facilities. Church building committees rarely have enough funds to build all the rooms and features they need. Therefore, we have often made provisions for the future installation of seating balconies, baptistry tanks, air-conditioning, speaker systems, stained-glass windows and other items the building budget could not allow at first.

Certain items, however, cannot easily be installed later, or would involve much greater cost at a later date. These include insulation in finished walls or ceilings, zoned heating, floor tiles (which need a *recessed* slab) and many other features. Your architect can advise you on the feasibility of deferring any items on your maybe list. However, you should think about the cost if you want to recess quarry tiles in an area where you have specified carpet on a concrete slab, for instance.

DO:
- Consider installing certain items and equipment at a future date and make provisions for such deferred installations.
- Understand that certain items cannot easily be installed later at the same cost, and therefore should not be deferred.

A part of "building better and for less" is avoiding pitfalls and foolish shortcuts. The next chapter will help you prevent your structure from becoming a museum of costly mistakes.

A-15 THE SEWAGE PUMP CURSE AND OTHER MEMORABILIA

When one of our neighbors built his house, he had two choices in connecting to the city sewer system. The sewer lines (mains) were not in our street but in the streets that run across it. He could connect at the nearest cross street, which, unfortunately, was uphill. This meant installing a sewage pump to push the sewage up the hill. Or he could have chosen to go downhill; but the distance to the next cross street was much longer, almost twice that going uphill. In addition to paying for the longer run, the sewage line would have to be tunneled under an existing alley. Also, refurbishing the neighbor's landscaped parking strip where the sewer line would have to be imbedded would be yet another expense. The shorter run uphill, including the sewage pump installation, did cost less than going the long way "down." Our neighbor chose the uphill run and has regretted it ever since.

Sewage pumps should be avoided, if possible. They are full of surprises even with the best pump and careful installation. The sewage pump curse is mentioned here as an example of an unfortunate solution which could have been avoided. In construction, certain equipment and certain materials can become lifelong headaches, impossible or difficult to correct (or only at disheartening costs).

In the foreword we promised not to sidetrack you to the drawing board to teach you how to design something—a skyscraper, a house, a deck, a backlighted bookshelf. We promised to show you how to get these things designed and bid, financed and built. But, you see, a part of "getting it built—better, faster, for less" is knowing how not to design and build it. A capable architect and experienced contractor will help you to avoid most building mistakes.

We all learn from each other. Therefore, on the following list are a few examples of follies we have observed or, in some cases, have made ourselves—or would have made, had not a knowledgeable contractor or supplier cautioned us. Like all lists in this book, this is not a complete collection of pitfalls, just a sampling.

In these examples from the main trades the **mistake** is in bold letters. What should have been done follows:

Earthwork: **Water seeps into building**
Drop outside concrete walks min. 1½ inches; soil: 6 inches minimum below sill plate; slope grade away from building for minimum 5 feet.

Concrete walks and patios:	**Water puddles** Slope ½ inch per foot.
Masonry retaining walls:	**Water rings show, paint peels off** Membrane—waterproof on earthside; install gravel and weep holes.
Carpentry:	**Wood floor squeaks** Use seasoned lumber; have short joist spans; use T&G plywood; lay in mastic; nail with sinkers at 6 inches on center.
Metal railings:	**Rust, even after being painted** Build in sections to fit into galvanizing tank. Galvanize completely before painting.
Exterior sliding doors:	**Rain or water from sprinkler seeps through sill** Caulk generously, per manufacturer's instructions, especially under sill and at corners. Immediately after installation, before exterior plastering, etc., make hose test.
High windows:	**Difficult to reach for window cleaning** Spot this common mistake in the planning stage; if windows cannot be reached by ladder or with a longhandle squeeze, provide a balcony or a ledge. Or ask your architect to clean the windows!
Drywall:	**Nail popping** Specify seasoned lumber, construction grade, not utility grade.
Sound insulation:	**Noise through bathroom walls** Install fiberglass insulation in walls around bathrooms; in addition, insulate pipe in those walls. In adjoining rooms, plan wardrobes along the walls that have bathrooms behind them.
Exterior painting:	**Paint blisters, bubbles, etc.** Instead of painting, stain as many exterior surfaces as possible.
Plumbing:	**Bathtub on wood floor settles** Increase joist size under tubs, space joists at 12 inches on center.
Heating:	**Heat is wasted** If possible, have warm air supplies installed in the floor and not in the ceiling or high up in walls. Consider installing a warm-air-recirculating fireplace. Consider two heating zones, with two smaller heating units instead of one large one. Two don't cost much more because each is a small model and you may save on supply and return ducts. Over the years, the double system can save a bundle of heating dollars. Heat won't be wasted in that portion of the house not used at a particular time, like the living and dining areas overnight for example.
Electrical:	**Difficult and costly to install across driveways, etc.** Install ½ inch pipes before pouring concrete for driveways and walks. The same goes for future low-voltage lighting wires—install conduits before pouring.
Sprinklers:	**Difficult to install after driveways and walks are in place** Install ½ inch or ¾ inch plastic pipes **before** pouring the concrete for driveways and walks.

The message is:

DO NOT install materials or equipment such as the sewage pump, the wrong exterior paint or difficult-to-reach windows where such an installation will become a headache or maintenance costs will drain your pocketbook for years to come.

Also invert this rule: DO INSTALL certain materials in a way better than standard construction even if the standard method is generally accepted and saves initially. The headaches come later: squeaking floors, sinking bathtubs on standard joist spacing or king-size heating bills with a standard heating system. Such mistakes cannot be easily corrected—sometimes they cannot be corrected at all. Catch these careless errors in the planning stage before they become memorabilia. The city fathers of Pisa can no longer add the missing reinforcing to the footings of their Leaning Tower!

Besides follies knowingly made or being penny-wise and pound-foolish, there can be hidden mistakes in drawings and specifications. The key is to spot them before someone builds them. A-16 tells you how.

A-16 MISTAKES AND DEFECTS IN DRAWINGS AND SPECIFICATIONS

There will be mistakes and defects in drawings and specifications, even some foolishness. For instance:

Ambiguity

The floor plan shows a symbol for a water heater, a circle with the letters "W.H." inscribed. But there should be another abbreviation to define the type of heater:

$$F.G. = \text{fuel gas}$$
$$220V = \text{electric}$$
$$P = \text{propane gas}$$

Why? The water heater will be furnished and installed by the plumber, and he must know what kind. But also, will (a) the plumber provide the gas line, or (b) the electrician install the 220 volt outlet, or (c) the propane gas company supply the tank and the line to the heater location? To get reliable bids, with no gaps and no overlapping, the contractors must know who furnishes what. Ambiguous information must be cleared up in the bidding stage.

Contradiction

The floor plan shows a water heater with an F.G. designation. The specifications, however, describe a 52-gallon electric water heater.

How serious are such mistakes? Not devastating, if it is one residential water heater. But any mistake costs the owner money, time and inconvenience, even if he is unaware of it. Ambiguities and contradictions are normally discovered either by the architect when checking over his drawings or by the plan checker in the Building Department or, in the bidding phase, by one of the contractors who prepare proposals. If drawings have already been sent out for bid by the time the mistake is discovered, the architect or owner should issue an addendum to all bidding contractors to clear up the fuzz. Could there be any other defects? You bet!

Wrong Design

For an apartment complex, the architect specified a group of water heaters as a central hot water system. The heaters did not produce enough water for the 58 apartments. If too many tenants had their dishwashers running while watching "All In the Family" others would not have enough water to take a shower—not even together! An untenable situation, right?

The hot water system should have been designed by a mechanical engineer or competent mechanical contractor, not an architect. But it was designed by an architect, and no one detected that the system was underdesigned until the building was occupied. The plan checker didn't catch it, nor the plumbing contractor. The architect was liable. Yes, he took a bath!

Omission
Four water heaters were located in a Utility Room on the second floor of an apartment building. The floor joists were not sufficient to carry the load of four full water heaters. Fortunately, the plan checker caught that one.

Foolishness
For a residence in a small rural town a natural gas water heater was specified, to be supplied from a ¾ inch gas meter. There is no natural gas supply in that town!

That's probably more chatter about water heaters than you ever wanted to hear. The heater, of course, is just an example. The point is, there can be ambiguities, contradictions, wrong design information, omissions and plain old foolishness in drawings or specifications.

Some years ago the San Diego Building Department concluded a study and issued a summary of major "Defects in Drawings Submitted to Plan Check." Sixty-two sets of drawings were analyzed. The survey did not list mistakes in specifications, since they are normally not checked by cities or counties. Nor did it mention defects in site engineering, electrical, mechanical or plumbing work. The survey was concerned with only the architectural and structural aspects. The major defects found in drawings for commercial buildings and apartments were:

Defect	% of drawings having this defect
Lateral load bearing capacity inadequate	52
Vertical load bearing capacity inadequate	48
Accommodation for the handicapped not provided	45
Inadequate fire resistive protection, exits, stairs	33
Inadequate occupancy separation	22

In 22 sets of drawings for single-family residences these defects were found:

Defect	% of drawings having this defect
Vertical load bearing capacity inadequate	41
Lateral load bearing capacity inadequate	32
Window area not sufficient	36
Fire-resistive construction inadequate	18

In general, what are the most frequent causes for lawsuits? There are three:
Stairs built incorrectly
Rain water leakage (roofs, balconies, doors, built-in downspouts)
Sagging of long span trusses

The message:
Mistakes in drawings and specifications can be costly if not cleared up before the construction contract is signed. For this reason, remember:

The contractor's bid must be based on approved drawings. No bid is realistic and complete if not based on drawings and specifications that have been approved by all governing agencies. As a practical matter, final bids cannot be received by the owner before the permits are issued, or are ready to be issued.

If bidding takes place before all governing agencies approve the drawings, copies of approved drawings should be sent again to contractors for review to update bids.

If the drawings are out for bid and changes to the drawings must be made on account of governing agencies, adjustments by the owner or modifications by the architect (such as the correction of mistakes), and the owner does not want to reroute the revised drawings to contractors, then the architect should issue to all bidding contractors an **Addendum,** which tabulates corrections and adjustments.

The construction contract will be based on the contractor's written proposal which will be based on drawings and specifications whether they are 60% or 99% correct.

How do you keep mistakes in drawings and specifications to a minimum?

DO:

- Hire competent professionals—architects, consultants, contractors.
- On large projects: Insist that the architect and his consultants (structural, mechanical, electrical, civil engineers) carry professional liability insurance.
- Have General Notes, including the following sentence in your contract documents, either on one drawing or in the specifications:

    The "General Conditions for the Contract for Construction" A.I.A. Document 201, dated (insert latest date) are herewith made a part of these notes. Copies of the document are on file with the Owner, Architect and General Contractor.

  The General Conditions urge contractors to carefully study and compare documents and to report to the architect any error, ambiguity or omission. Seldom will the contractor be held liable for mistakes in drawings and specifications, but at least the above sentence obligates the contractor to study all the documents: therefore, mistakes can come to the fore.

- Issue addenda to all bidding contractors to clear up mistakes and errors before the bid opening.
- Have in your contracts with architects, consultants and contractors a clause that makes arbitration mandatory. If mistakes in construction documents cause legal problems, arbitration costs less and is faster than litigation. If in doubt, have an attorney look over the contract documents before you sign them.

THE AMERICAN INSTITUTE OF ARCHITECTS

AIA Document A201

General Conditions of the Contract for Construction

THIS DOCUMENT HAS IMPORTANT LEGAL CONSEQUENCES; CONSULTATION WITH AN ATTORNEY IS ENCOURAGED WITH RESPECT TO ITS MODIFICATION

1976 EDITION
TABLE OF ARTICLES

This document has been approved and endorsed by The Associated General Contractors of America.

BUILDING INSPECTION DEPARTMENT
PLAN CORRECTION SHEET
APARTMENTS, HOTELS, MOTELS AND GARAGES
THIS IS NOT A BUILDING PERMIT

THE FOLLOWING LIST DOES NOT NECESSARILY INCLUDE ALL ERRORS AND OMISSIONS. SEE SEC. 302(C) OF THE UNIFORM BUILDING CODE. PLANS REQUIRE CORRECTION AS INDICATED BY CIRCLED ITEMS BELOW BEFORE A BUILDING PERMIT CAN BE ISSUED. RETURN THIS CORRECTION SHEET WITH CORRECTED PLANS. TO FACILITATE RECHECKING PLEASE IDENTIFY, NEXT TO EACH CIRCLED ITEM BELOW, THE SHEET OF THE PLANS UPON WHICH THE CORRECTION HAS BEEN MADE.

CLEARANCE
NOISE ABATEM
COAST REGIC
ELECTRICA'
PLUMBING
HEALTH
FIRE PF
ENGIN
UTILI
OLD
PRO
POI
F.
7

A. Permit Application
1. Permit application shall be signed by the owner or his authorized agent. (Sec. 301)
2. Provi'

D. Exits
1. Two

A-17 WHILE THE DRAWINGS ARE BEING CHECKED BY THE BUILDING DEPARTMENT

Your drawings will be at the building department for some time:

Residences, including remodelings	2 to 4 weeks
Apartments, commercial projects, subdivisions	3 to 6 weeks
The blockbusters: highrises, hospitals, schools	4 to 12 weeks

The time depends on how many sets of drawings are submitted, the department's efficiency, which varies from city to city and the workload, which varies from doldrums to pandemonium. Let's see how you can make good use of the time during which the drawings, specifications, soil reports, structural calculations, affidavits and other forms are waiting for approval in the building department. "Building department" meaning all the governing agencies that might check your submitted project:

Civil engineering division
Architectural and structural plan checkers
Electrical, mechanical and plumbing engineers
Utilities division
Zoning

and in some cases:

Coastal Commission
Health Department
Fire Marshal
Port Director
Civil Aeronautics Administration

Will the wait be six or twelve weeks? Ask this question when the drawings are submitted. Two sets of drawings with one copy each of specifications, reports and calculations is the minimum requirement. More sets can speed up the approval process. If you think your case is an emergency, ask to see the chief plan checker. Some building departments have a "Quick Bin" for plan checking.

And then, don't just sit there—use the time.

► Get yourself a complete set of prints of the drawings and a set of specifications as submitted for plan check and read every sheet—every number, every reference note, every general note. If you want to be close to perfect and as knowledgeable (almost) as your architect or draftsman, take three colored pencils—red, blue and yellow—and cross every number and note as follows:

Red—what you think is wrong
Blue—what you question or don't understand
Yellow—what you agree with.

If you want to get an A+, do the same with specifications.

When the drawings come back from the building department for "Building Department Corrections" confer with your architect about your red and blue marks.

Why would you want to spend your Sundays with three colored pencils? The careful line-by-line, critical reading of your future contract documents makes you familiar with the building like no other exercise can. It gives you a chance to make last minute adjustments before the permit is issued and prepares you for your job visits during construction.

▶ After this check-reading you or your architect should give one set of drawings and specifications to one of the general contractors who later will be invited to bid. Ask him to tentatively estimate the cost range of the project. (No contractor can give you a guaranteed cost at this stage. See A-16.) If you act as owner-builder, without hiring a general contractor—and we advise strongly against it except for the smallest project—in order to get a tentative cost range you will have to give prints and specifications to all your subcontractors, at this time to one of each trade. Most subcontractors want a complete set of drawings, or at least all the drawings and specifications that affect their work. The electrician, for instance, needs not only the floor plan to see where meter, outlets and panels are, but the site plan, to see how to get his conduits and wires from a transformer or pole to your meter location. He also needs sections, to see the heights of rooms, and the plumbing fixture schedule, so he will know that he has to connect the garbage disposal and other equipment.

Upon receiving the one general contractor's tentative cost range, compare it with the architect's estimate of "probable construction costs." Arrange a meeting with him and the contractor to talk about costs. Do you see a need to downgrade, upgrade or delete at this stage?

▶ If pressed for the building's completion time, you could start bidding with all contractors while the drawings are in the city or county departments. The disadvantage is, as mentioned before: When drawings come back from the building department for "corrections" these adjustments must be made known to all bidders by an addendum. Otherwise, the bids will not be complete. And maybe you or your architect want to make last-minute adjustments. These also have to be explained to all bidders by way of addendum. Later, after all adjustments have been made on the original drawings, sets must be printed for the construction phase. And that means plenty of new sets, probably between 15 and 25—another additional expense to the client.

The message: If you are not pressed for time, don't ask for final bids until drawings and specifications are signed off by all governing authorities and the architect.

▶ If you have not done it before, now is the time to visit showrooms to select and confirm materials, such as entrance doors, finish hardware, flooring, plumbing fixtures, windows, ceramic tiles, etc. Your drawings call out these items. The specifications, let's hope, nail down the type and quality level. Within that quality stratum you can change from one brand to another without affecting the building department check or any bidding already underway. Keep in mind for later: Even in the construction phase, you can change your mind on specified materials, as long as the materials are of the same quality level and as long as no materials have been ordered by suppliers. If you change **after** materials have been ordered by contractors, or from one type to another (sheet vinyl to carpet, for instance), or if you change the quality level (regular sheet) vinyl to cushioned sheet vinyl), then the plot thickens. Read C-10.

▶ Meet with prospective lenders and put the last touches on lining up the construction financing. Funds must be assured before you sign the construction contract and begin construction. Fill in tentatively the construction cost breakdown form. Remember to plug in not only the construction costs, but all incidental costs as well. Read A-9 again. While your drawings are being checked by the Building Department:

DO:
- Review all construction documents.
- Verify materials specified.
- Complete details, color schedules and other aspects of drawings not required by the Building Inspection Department in this period. (This note is directed to the owner who produces his own drawings.)
- Clear your drawings and specifications with the Architectural Control Board.
- Tie up financing in the form of a written commitment from the lender.
- Monitor the processing of all special permits. They must be processed and in place by the time you clear for all permits or when you want to take out the building permit.

The following sketch shows in a comprehensive and somewhat confusing way how to obtain all permits. But it is a true picture: the process can be confusing.

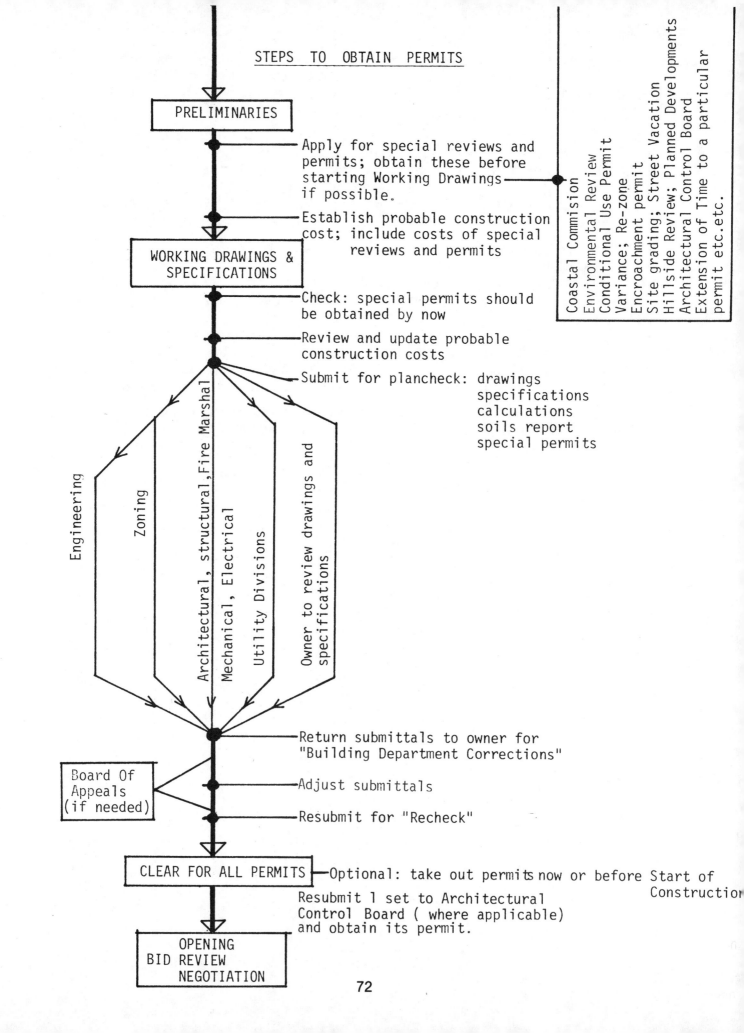

STEPS TO OBTAIN PERMITS

PRELIMINARIES

Apply for special reviews and permits; obtain these before starting Working Drawings if possible.

Establish probable construction cost; include costs of special reviews and permits

WORKING DRAWINGS & SPECIFICATIONS

Check: special permits should be obtained by now

Review and update probable construction costs

Submit for plancheck: drawings
specifications
calculations
soils report
special permits

Coastal Commision
Enviromental Review
Conditional Use Permit
Variance; Re-zone
Encroachment permit
Site grading; Street Vacation
Hillside Review; Planned Developments
Architectural Control Board
Extension of Time to a particular permit etc.etc.

Engineering

Zoning

Architectural, structural, Fire Marshal

Mechanical, Electrical

Utility Divisions

Owner to review drawings and specifications

Return submittals to owner for "Building Department Corrections"

Board Of Appeals (if needed)

Adjust submittals

Resubmit for "Recheck"

CLEAR FOR ALL PERMITS

Optional: take out permits now or before Start of Construction

Resubmit 1 set to Architectural Control Board (where applicable) and obtain its permit.

OPENING
BID REVIEW
NEGOTIATION

72

A-18 HOW LONG DOES ALL THIS TAKE?

Planning and permits, bidding and financing, selecting a contractor, and then—construction. In most cases, it takes 5 to 12 months. Wow! That is longer than most owners anticipate. But it takes time to go through all these steps:

Procedure	Minimum time	Probable time
Confer with architect or draftsman; assemble information; formulate tentative building program (3 lists)	1 week	2 weeks
Get survey and soil information; produce preliminaries; meetings with architect; produce construction documents; clear for all permits. (Allow more time for Special Reviews and Permits)	4 weeks	9 weeks
Obtain and evaluate bids; obtain loan commitment, select contractor; receive Notice to Proceed from lender; sign owner/contractor agreement.	3 weeks	5 weeks
So far that could be from:	8 weeks	16 weeks
This applies to small to medium projects (residences, apartments, commercial structures). Projects in coastal zones or with soil problems, planned residential developments, subdivisions, large buildings and projects under V.A. or F.H.A. may take much longer.		
Then comes the construction phase:	12 weeks to	36 weeks
A total of about	20 weeks to **5 months**	52 weeks **12 months**

Can the total process be accelerated? Yes, by doing certain work simultaneously or by some overlapping instead of following a strict step-after-step sequence. The building trade calls this method "fast-tracking":

▶ As early as possible have the architect involved in site selection or evaluation and order surveys and soils reports. If special permits are required, let him apply for them immediately.

As soon as preliminary drawings have begun, select a good general contractor and arrange with him a "cost-of-the-work-plus-fee contract." With larger commercial projects the architect's consultants and electrical and mechanical contractors should also be on board early. Upon tentative approval of preliminary drawings by the client, the general contractor should start estimating costs. He will update his estimates through the working drawing stage all the way down the line until the construction agreement is signed.

▶ Line up financing as soon as preliminaries are available. Shop aggressively for the best terms. The lender's package of preliminary drawings and specifications should *look* as complete as possible. Besides site plan, floor plans and elevations, they should include preliminary sections, schedules, a "General Notes and Standard Detail Sheet" and preliminary specifications on the FHA type "List of Materials."

▶ If an early construction start is mandatory, obtain a foundation permit and begin building before the final permit is issued—this method can save you weeks. Sometimes such an early start gets your project ahead of bad weather or even affords more favorable financing in times of rising interest rates. The foundation permit drawback: Once foundations are poured, the building layout is almost frozen in regard to size, shape, load bearing walls, column locations and shear walls. This is normally not a problem, however. The time lag between a foundation permit and a final permit is about 4 to 8 weeks. Most buildings are not changed dramatically during these last few weeks of the planning phase.

▶ In regard to bidding, the early start with a foundation permit necessitates that you abandon the one rule we stressed before: to have drawings and specifications completed and signed off before bidding. Ask the general contractor to start bidding the project when the architect **submits** drawings and specifications to the building department. Later, the architect will update all drawings on account of building department corrections, client adjustments and his own corrections. As soon as the permits are issued and the architect has signed off all documents "For Construction" the general contractor needs from 15 to 25 sets of updated prints of drawings and specifications and must issue these again, with an addenda, to all bidding subcontractors in order to update his contract bid.

▶ Even if you "fast-track" it is important to have financing secured and the Notice to Proceed received from the lender before the first shovel of dirt is turned. When starting with just a foundation permit, the lender must understand that the exact building costs cannot be known when the loan is negotiated, since your drawings are still in the building department with the main portion of your drawings being checked (or sometimes, still being designed and not even in plan check!)

Fast-tracking — it's no picnic!

DO:
● Allow ample time for planning. It is cheaper to move a wall on the drawing than on the construction site.

- Allocate plenty of time for bidding, shopping for financing, waiting for approval by all governing agencies.
- Resubmit drawings and specifications to architectural control board, where required. (See A-11.)
- "Fast-track" if necessary. Take advantage of the experience of both architect and contractor. Expert professional help is a must.

Repeat Note: Getting something built often takes longer than even the experts estimate. The flow chart on the next page gives you an overview.

Can you build faster, avoid the hassle with architects, building departments, OSHA, Coastal Commission and inspectors? Can you save the fees for reports and permits, by just building without permits? Verrry interesting! Something you always wanted to know but were afraid to ask.

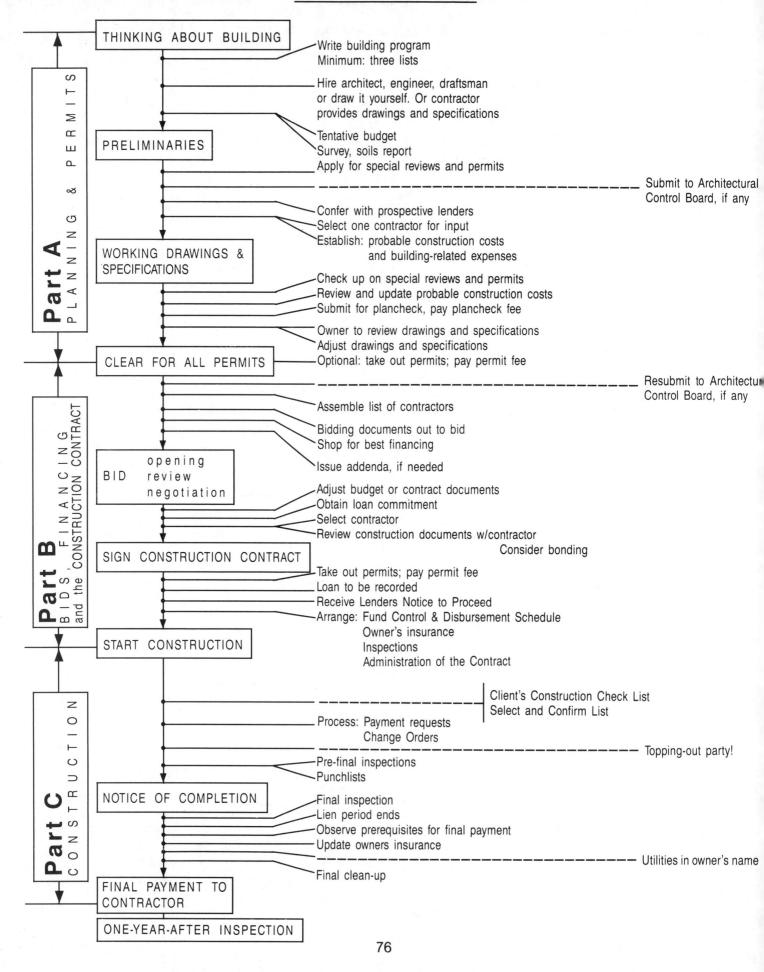

Part A — PLANNING & PERMITS

THINKING ABOUT BUILDING
- Write building program
  Minimum: three lists
- Hire architect, engineer, draftsman
  or draw it yourself. Or contractor
  provides drawings and specifications

PRELIMINARIES
- Tentative budget
- Survey, soils report
- Apply for special reviews and permits
- - - - - - - - - - - Submit to Architectural
  Control Board, if any
- Confer with prospective lenders
- Select one contractor for input
- Establish: probable construction costs
  and building-related expenses

WORKING DRAWINGS & SPECIFICATIONS
- Check up on special reviews and permits
- Review and update probable construction costs
- Submit for plancheck, pay plancheck fee
- Owner to review drawings and specifications
- Adjust drawings and specifications

CLEAR FOR ALL PERMITS
- Optional: take out permits; pay permit fee
- - - - - - - - - - - Resubmit to Architectural
  Control Board, if any
- Assemble list of contractors
- Bidding documents out to bid
- Shop for best financing
- Issue addenda, if needed

Part B — BIDS, FINANCING and the CONSTRUCTION CONTRACT

BID — opening review negotiation
- Adjust budget or contract documents
- Obtain loan commitment
- Select contractor
- Review construction documents w/contractor
  Consider bonding

SIGN CONSTRUCTION CONTRACT
- Take out permits; pay permit fee
- Loan to be recorded
- Receive Lenders Notice to Proceed
- Arrange: Fund Control & Disbursement Schedule
  Owner's insurance
  Inspections
  Administration of the Contract

START CONSTRUCTION

Part C — CONSTRUCTION
- Client's Construction Check List
  Select and Confirm List
- Process: Payment requests
  Change Orders
- - - - - - - - - - - Topping-out party!
- Pre-final inspections
- Punchlists

NOTICE OF COMPLETION
- Final inspection
- Lien period ends
- Observe prerequisites for final payment
- Update owners insurance
- - - - - - - - - - - Utilities in owner's name
- Final clean-up

FINAL PAYMENT TO CONTRACTOR

ONE-YEAR-AFTER INSPECTION

A-19 BUILDING WITHOUT A PERMIT

Did you hear about the friend of a neighbor who added a carport and converted the garage into a family room—without a permit—and got away with it? Is it smart to build without the required general and special permits?

When someone builds without a permit, he has to consider **two** aspects—**legal** and **practical**. Let's look at the legal side first. From the Uniform Building Code, which governs most building activities in the Western states:

> No person, firm or corporation shall erect, construct, enlarge, alter, repair, move, improve, remove, convert or demolish any building or structure in the city, or cause the same to be done, without first obtaining a separate building permit for each such building or structure from the Building Official . . .
>
> Anyone who erects, constructs, etc., without a permit will have to pay double the required fee . . .
>
> . . . but the payment of such double fee shall not relieve any person from fully complying with the requirement of the Uniform Building Code in the execution of the work, nor from any other penalties . . ."

That last sentence can be the real blow if the owner gets caught. Why? Let's assume the plan check and permit fee for the carport and garage conversion would have been $70. But the man didn't take out a permit. So, Mr. Nopermit had to pay double the fee: He swallows that. But what if the carport does not "fully comply" with the requirements of the code, perhaps the zoning code? What if the carport was erected unlawfully within the required minimum sideyard, too close to the property line? No permit can be obtained since the owner cannot "comply fully with the requirements of the Uniform Building Code." He may get a variance, but it's not likely. The carport has to come down, even though he is willing to pay the fee twice. It has to come down and that costs money. He will live on Rolaids for the next month.

Summing up, what are the gains (+) by building without the necessary permits?

+ No expenses for drawings, plan check and permit fees.
+ No copy of the permit will be forwarded to the tax assessors, therefore property taxes will not increase correspondingly with the added value of carport and family room, unless of course, the tax man stumbles onto the project during construction.

Then, what are the risks (−)?

− Being caught and having to pay double the fee—a small risk with a small project.

– Being caught and finding out that the halfway or fully completed project cannot be permitted because it cannot comply with the codes. You still have to submit a full set of drawings and pay a double plan check fee. Then you might be informed that a permit cannot be issued. You might try to obtain a variance; it could and probably would be denied.

    The Taj Mahal has to come down. The costs: useless drawings, lost fees, expensive labor, first to build then to remove the thing, wasted materials. If the zoning administrator is tough enough he makes you restore the garage to provide "proper off-street parking." Truly a nightmare. In most cases, it's not worth the risk.

Then there is the practical disadvantage of building without permits:

– Not getting the protection value a permit offers, the free professional inspections by the city or county. Even the smallest project, like the carport-garage conversion will be inspected at least three times: (1) after excavation, and before footings are poured: (2) after framing and before walls are closed; and (3) after substantial completion.

    In addition to the three inspections, larger projects have more visits by inspectors: grading, drywall, electrical, heating, venting, plumbing, final grading and others. So why not get the benefit of the inspectors' experience and their opinions on the work performed, not to mention the mistakes they certainly will discover and insist on being corrected? (Some contractors and workmen perform better, too, when they know there will be inspections.)

    True, city inspectors are officially concerned with only two aspects of your project:

• Compliance with engineering and zoning regulations.

• Enforcement of Uniform Building Code requirements, which deal with safety, health and, as of late, energy-saving requirements.

    Safety and health—the city inspector's bag. Not beauty, serenity, more trees, less concrete, old world workmanship, pride in work, how to get the guy back to finish the roofing. Yet, when an inspector pokes around to insure that safety and health regulations are enforced, including such minute details as adequate nailing and spacing of ceiling joists, he can, beyond the call of duty, give you an opinion on other matters—workmanship, material grades, subcontractors' reputations and other useful information. This means that without official inspections you not only miss a reliable source of seasoned experience; worse, you will have little protection against unscrupulous contractors who grace you with shallow footings, undersized joists, sloppy roofing, faulty wiring.

    Needless to say, no architect or building designer of sound mind will oversee a project for which a permit is required but not obtained. Remember: ". . . no person . . . shall cause . . . to alter, repair, improve"! That includes the architect!

    One more wrinkle: There could be a situation similar to "building without a permit": the permit was obtained but has expired.

    It happens in one of two variations.

► The **permit application expires** and the plan check fee will be forfeited if the submitted drawings are not picked up again by the owner or architect within 180 days of submittal. Some Building Officials grant an extension if an application for extension is made **before** the application expires; others just coldheartedly charge a new plan check fee.

    or

► After the permit is issued, the building permit will **automatically expire** if either (a) the first site inspection was not conducted within 120 days from permit issue date, or (b) later on, the work on the project ceases for a period of more than 120 days. In either case the permit will be void automatically; "by statute" they call it. No building official can grant an extension. The owner can revive the permit by paying again **one-half** the permit fee.

The matter can be further complicated: About every three years, a new building code is adopted. If you diddle around and let your permit expire at about the time the building code expires the plot thickens considerably—some building departments will not renew the permit. The entire project must be rechecked against the new code.

DO:

- Check whether a permit is required for your project.
- Take out permits, IF they are required.
- Mark with red on your calender: 180 days allowed to pick up drawings from the building department and resubmit for permitting; 120 days from day of permit to the first field inspection. Verify these limits with your local Building Department.

You have read unflaggingly to here: all hardhats off to you! Now study the Cost-Shaving List on the following page. Check your project against every recommendation.

Refer to	DO	Importance to reduce construction costs, hazards or liabilities	✓ Pertains to your project	✓ Done
A-3	Set up a separate bank account for the building project. Record all expenses. Keep receipts to offset capital gain taxes at a future date.	■		
A-1	Arrange for a PRELIMINARY CHECK with: Zoning and Building Department, Engineering Dept. Sometimes: Fire Marshal, Health Dept., Review Boards. With commercial projects: check with future tenants.	■ ■ / ■		
A-11	Review/verify: Deed restrictions, easements, Soil conditions, earthquake vulnerability, Covenants, Conditions & Restrictions, Utilities available or adequate, Need for sound control, Provisions for handicapped persons.	■ / ■ / ■ ■ / ■ ■ / ■ / ■ / ■		
A-6	Have contract documents suitable for competitive bidding And yet: don't overdraw. And streamline specifications.	■ ■ ■ ■		
A-7	Consider this fee arrangement with architect etc.: Preliminaries and Contract Administration: Hourly rate Contract documents: fixed fee	■ ■		
A-13, A-5	Give preference to: standard materials, equipment, sizes, colors. Locally Available.	■ ■		
A-6	Include as many energy-saving features as your budget allows.	■ ■		
A-16	Include A.I.A. General Conditions and your Special Conditions. State that these conditions govern in case they are in conflict with contractors' proposal forms.	■ ■		
	Include protective General Notes.	■		
A-9	Anticipate and scrutinize building-related expenses: survey, soils report, financing, review board fees, printing, permits, bonds, professional fees, fund control, insurance, utilities during construction, testing, special assessments. Consider to include many of these in the construction loan.	■ ■ / ■ ■		
A-4	Expect higher cost when building under F.H.A. or V.A.	■		
A-14	Prepare a list of lower-priced alternates or "provisions for future installations."	■ ■		
A-17	Be aware of the cost of changing drawings.	■		
A-18	Fast-track where feasible.	■		
B-14	Date each drawing, print, specification, addendum, change order.	■ ■ ■		
B-14	Verify cost of permits when taken out by the contractor.	■		
A-19	Read scope and expiration date of permits.	■		
Refer to	**DON'T**			
A-7	Guarantee a contractor the job in exchange for free drawings.	■ ■		
A-7	Be a structural hero.	■ ■		
A-19	Build without required permits.	■ ■		

CITY OF SAN DIEGO BUILDING PERMIT APPLICATION

LEGAL DESCRIPT'N

ADDRESS		SUITE
LOT	BLOCK SUBDIVISION NAME	UNIT NO.

EXISTING USE	PROPOSED USE

NAME	TEL. NO.
ADDRESS	ZIP CODE

NAME	
ADDRESS	TEL. NO.
CITY	ZIP CODE
STATE LIC. NO.	LIC. CLASS	CITY LIC. NUMBER

NAME	LIC. NO.
ADDRESS	TEL. NO.
CITY	ZIP CODE

LICENSED CONTRACTORS DECLARATION: I hereby affirm that I am licensed under provisions of Chapter 9 (commencing with Section 7000) of Division 3 of the Business and Professions Code, and my license is in full force and effect

Contractor's Signature

OWNER-BUILDER DECLARATION: I hereby affirm that I am exempt from the Contractor's License Law for the following reason (Sec 7031.5, Business and Professions Code Any city or county which requires a permit to construct, alter, improve, demolish, or repair any structure, prior to its issuance, also requires the applicant for such permit to file a signed statement that he is licensed pursuant to the provisions of the Contractor's License Law (Chapter 9 (commencing with Section 7000) of Division 3 of the Business and Professions Code) or that he is exempt therefrom and the basis for the alleged exemption Any violation of Section 7031.5 by any applicant for a permit subjects the applicant to a civil penalty of not more than five hundred dollars ($500))

☐ I, as owner of the property, or my employees with wages as their sole compensation, will do the work, and the structure is not intended or offered for sale (Sec 7044, Business and Professions Code The Contractor's License Law does not apply to an owner of property who builds or improves thereon, and who does such work himself or through his own employees, provided that such improvements are not intended or offered for sale If, however, the building or improvement is sold within one year of completion, the owner-builder will have the burden of proving that he did not build or improve for the purpose of sale)

☐ I, as owner of the property, am exclusively contracting with licensed contractors to construct the project (Sec 7044, Business and Professions Code The Contractor's License Law does not apply to an owner of property who builds or improves thereon, and contracts for such projects with a contractor(s) licensed pursuant to the Contractor's License Law)

☐ I am exempt under Sec _____ B &P C for this reason

Owner's Signature

WORKER'S COMPENSATION DECLARATION: I hereby affirm that I have a certificate of consent to self-insure, or a certificate of Workers' Compensation Insurance, or a certified copy thereof filed with the Building Inspection Department (Sec 3800, Lab C)

Policy No _____ Insurance Company _____

Applicant's Signature _____ Expiration Date

CERTIFICATE OF EXEMPTION FROM WORKERS' COMPENSATION INSURANCE: I certify that in the performance of the work for which this permit is issued, I shall not employ any person in any manner so as to become subject to the Workers' Compensation Laws of California

Owner's or Contractor's Signature

CONSTRUCTION LENDING AGENCY: I hereby affirm that there is a construction lending agency for the performance of the work for which this permit is issued (Sec 3097, Civ C)

Lender's Name _____

Lender's Address

NAME		
ADDRESS	CITY	ZIP CODE

I certify that I have read this application and state that the above information is correct, and that I am the owner or the duly authorized agent of the owner. I agree to comply with all city and state laws relating to building construction. I hereby authorize representatives of the City of San Diego to enter upon the above-mentioned property for inspection purposes. If, after making the Certificate of Exemption, from the Worker's Compensation provisions of the Labor Code I should become subject to such provisions, I will forthwith comply. In the event I do not comply with the Workmen's Compensation law, this permit shall be deemed revoked.

SIGNATURE _____ DATE _____

☐ CONTRACTOR ☐ OWNER ☐ AGENT FOR CONTRACTOR ☐ AGENT FOR OWNER

258 Rev. 4-85

PROJECT ADDRESS		SUITE
DIST. NO.	CENSUS TRACT NO.	PERMIT NUMBER

CONDITION OF SOIL AT PROJECT
☐ UNDISTURBED
☐ COMPACTED FILL
☐ LOOSE FILL

COORD. INDEX NO.	PLAN FILE NO.

DESCRIPTION OF PROPOSED WORK _____

BUILDING INSPECTION DEPARTMENT

THIS PERMIT AUTHORIZES ONLY THE WORK NOTED

CITY TREASURER VALIDATION

SERV. SIZE	METER SIZE	CREDIT		WATER	SEWER
SEWER CONN.	☐ SEE ATT. RECPT.		CKD BY.		

USE ZONE	COASTAL ZONE ☐ YES ☐ NO	DEV. PERMIT NO:	NO. OF BAR SINKS	B.C. CODE	DWELL UNITS
	HILLSIDE REVIEW ☐ YES ☐ NO	PLANNED DISTRICT ☐ YES ☐ NO	BEDROOMS	S 1 2	
	AGREEMT. NO.	VARIANCE NO.		3 4	

FUND & ACC'T.	VALUATION OF WORK ▸	NO. UNITS	PER UNIT	ACCOUNT ITEM TOTAL
41300 73421	PLAN CHECK FEE			
	SUPPL. PLAN CHECK FEE			
41300 73422	BUILDING PERMIT FEE		*	
63010 9022	STATE FEE			
41506 77565	SEWER FEE			
41500 77530	WATER FEE			
73423	PARK FEE			
HOLD ORDER NO.				

*PENALTY FEE(S) ADDED AS PROVIDED BY MUNICIPAL CODE

WORK TO BE DONE
☐ NEW ☐ ALTER ☐ MOVE
☐ ADD'N. ☐ REPAIR ☐ DEMO.
☐ RES. ☐ K-RES. ☐ OCC. CH.

TOTAL FEES DUE ▸

TYPE CONST'N	NO. STORIES	OCCUP. GROUP
BUILDING AREA		TOTAL FLOOR AREA

SPEC. INSP. REQ'D. FOR
☐ CONCRETE ☐ MASONRY
☐ WELD. HS BOLTS ☐ SOILS
☐ PILE DRIVE ☐ OTHER

SPRINKLERS REQD FOR

PLANS CHECKED	DATE
PLANS APPROVED	DATE

DATE PLANS SUBMITTED

PLAN CHECK RECEIPT NO.

PLAN CHECK REC'PT. AMT. $

APPLICATION APPROVAL
SIGNATURE OF BUILD. INSP. DEPT. DEPUTY

MECH. PLAN CHECKER SIGNATURE RELATED TO TITLE 24 REQ'RMTS.

OCCUP. CARD PREPARED BY PLAN CHK. YES ☐ NO ☐

DATE:

INSPECTOR

81

PART B: BIDS, FINANCING AND THE CONSTRUCTION CONTRACT

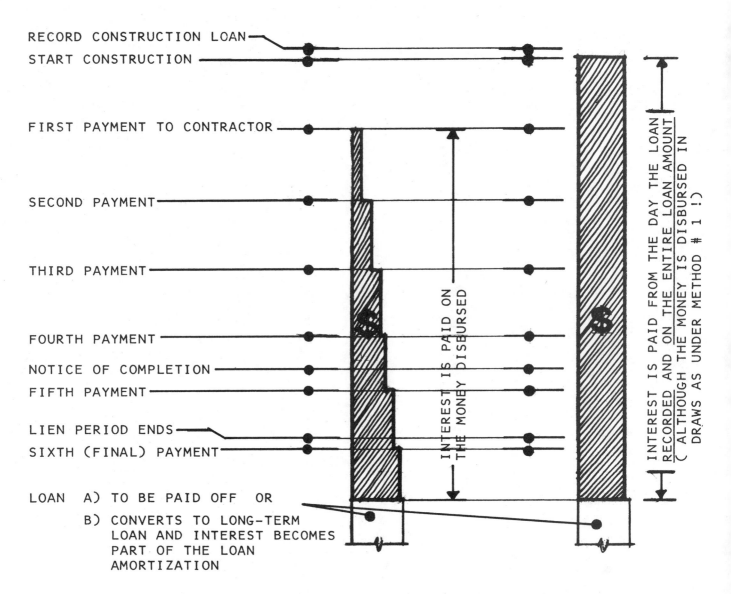

RECORD CONSTRUCTION LOAN
START CONSTRUCTION

FIRST PAYMENT TO CONTRACTOR

SECOND PAYMENT

THIRD PAYMENT

FOURTH PAYMENT
NOTICE OF COMPLETION
FIFTH PAYMENT

LIEN PERIOD ENDS
SIXTH (FINAL) PAYMENT

LOAN A) TO BE PAID OFF OR
B) CONVERTS TO LONG-TERM LOAN AND INTEREST BECOMES PART OF THE LOAN AMORTIZATION

INTEREST IS PAID ON THE MONEY DISBURSED

INTEREST IS PAID FROM THE DAY THE LOAN RECORDED AND ON THE ENTIRE LOAN AMOUNT (ALTHOUGH THE MONEY IS DISBURSED IN DRAWS AS UNDER METHOD # 1 !)

AMOUNT OF INTEREST ON THE CONSTRUCTION LOAN

1

WHEN INTEREST IS PAID ON THE MONEY DISBURSED

2

WHEN "DUTCH INTEREST" IS PAID

This illustration shows how much more interest you would pay if it were not "interest on the money disbursed" but "dutch interest."

B-1 WHICH COMES FIRST: BIDS OR FINANCING?

Fortunately, there is a logical sequence to the building process—the time sequence.

| YOU DECIDE TO BUILD |

write a building program (3 lists)

| get PRELIMINARIES |

establish a tentative budget
if needed: obtain a survey, soils report
and apply for special permits
confer with prospective lenders

START TALKING WITH LENDERS

| proceed with
WORKING DRAWINGS and
SPECIFICATIONS |

submit for plancheck
review: drawings, specifications, budget;
comply with Building Department corrections

KEEP TALKING— LOOK FOR BEST TERMS!

| CLEAR FOR ALL PERMITS |

invite contractors to bid

| REVIEW BIDS |

adjust contract documents, budget;
obtain loan commitment
select contractor

SELECT BEST LENDER

| SIGN THE CONSTRUCTION AGREEMENT
the "contract" |

take out permits
receive lender's Notice to Proceed

| START CONSTRUCTION |

You noticed that this diagram is a part of the flow chart of a typical building project shown in Chapter A-18.

At what point in time do you get a building loan? Should it be done before proceeding with working drawings to make sure you can obtain the loan? Or should you wait to talk with lenders until the contractors' bids are in so you can inform the lender about the exact building costs and how much you want to borrow? By that time you will have paid for complete drawings and specifications, plancheck and maybe special permit application fees. If you cannot get the loan, all these expenses will have been for nothing.

For these reasons it is best to start talking with lenders early in the planning stage. Establish a probable construction cost range (by the cost-per-square foot method) soon after you decide to build.

Let's use a residential project as an example:

Land costs		$ 30,000
Probable construction costs (including contingency allowance)		98,000
Building-related costs per A-9		15,000
	TOTAL	$ 143,000
Let's assume the lot is paid for.	Deduct	30,000
You intend to acquire a 2nd Trust Deed, with no payments for 2 years	Deduct	30,000
So you need to borrow		$ 83,000

There are two main types of loans to consider:

1. A CONSTRUCTION LOAN provides the funds for construction for a certain period, from 6 to 12 months. It must then be paid off by a permanent loan, which either you—or the next buyer, in case you sell—has to obtain.

2. A combination CONSTRUCTION/PERMANENT LOAN, which initially provides $83,000 for construction and after 6 to 12 months, after completion of construction, converts to permanent financing. You start paying off the $83,000 in equal monthly installments over the next 25, 29½ or 30 years. Terms differ from lender to lender, some might have even variable interest rates or participation in appreciation.

    If the CONSTRUCTION/PERMANENT LOAN is up to, say, $40,000 and will be used to remodel or rehabilitate an existing dwelling, the money could be obtained by a HOME IMPROVEMENT LOAN, to be paid back in 15 years or less. While the typical construction or construction/permanent loan is secured by a first trust deed, home improvement loans can be made in form of second trust deeds.

Some lenders grant EQUITY LOANS, also up to about $40,000 or $50,000, which are based on the existing equity in real estate used for dwelling purposes; business properties are normally not eligible. The proceeds may be used for almost any purpose; a second trust deed secures the loan. Equity loans differ from Home Improvement Loans in two ways: they are costlier and the lender is not really interested in for what purpose the loan is finally used. No rigmarole with drawings, cost breakdowns and lenders' inspections.

To secure either type of loan, talk with lenders early since they need four to six weeks to process the application. Be sure the loan is recorded before the start of construction. Lenders understand that exact building costs cannot be obtained without completed drawings and specifications, which are available only after drawings are cleared for permits. Therefore, most lenders are satisfied with "progress drawings and specifications"

86

and a statement of probable construction and building-related costs. After permits are issued and before construction begins, however, the lender is entitled to receive a copy of the approved drawings and specifications.

Here is an approximate timetable for arranging financing: As soon as you have preliminary drawings, know the probable construction cost and how much you want to borrow, make appointments with lending officers and present the following documents:

- preliminary plot plan
- floor plan with finish schedule
- exterior elevations
- outline specifications (use F.H.A. type "List Of Materials", Page 32)
- cost estimate based on cost per square foot
- up-to-date financial statement
- building loan application (forms are available at any savings and loan association). Don't hesitate to use the application form of any lender to present to any other lender. Eventually, the lender will ask you to use his forms, but for a first meeting any lender's form will do.

How do you find out which lender is in the market for a loan? The escrow department of your bank is a good source. Also query loan officers of savings and loan associations, contractors, fund control departments, friends who have built, architects, classified ads and Chicago Title Insurance Company's "Regional Loan Guide."

When meeting with the loan officers, carefully make notes of the terms of each lender:

- type of loan: construction, construction/permanent, permanent only, equity
- amount you can borrow
- does it include a land draw? (if needed)
- loan fee and other costs
- prepayment penalties
- interest rate, in annual percentage rate (APR)
- does lender want (a) fund control? (b) a title insurance policy for the loan?
- length of loan—choose as long a term as possible.
- what documents must owner furnish with loan application? Ask the lender for a "Construction Loan Checklist" to know exactly what papers to submit.
- any objection to a Second Trust Deed, in case you need one?
- assumption rights

Could your request for a loan be denied? It is revealing to read a checklist used by lenders to determine the weak points in an applicant. Study the list. It gives these reasons:

Credit application incomplete	Insufficient credit file
Length of employment	No credit file
Insufficient income	Excessive obligations
Insufficient credit references	Delinquent credit obligations
Unable to verify credit references	Unable to verify income
Bankruptcy	Inadequate collateral
Too short a period of residence	Temporary residence
Unable to verify residence	Garnishment, attachment, fore-
Unable to verify employment	closure suit or repossession

Or you might find this straightforward reply to your request: *"We do not grant credit to your application on the terms you requested."* Did you ask for a 50-year loan, at 6½ percent, with no points and no payments until Christmas? Remember the Golden Rule:

He who has the gold makes the rules.

Tentatively select a lender with the most favorable terms and try to get a written commitment spelling out the loan terms in detail. As soon as you've selected a contractor, submit to your lender a set of drawings and specifications as approved by the building department, an updated financial statement, a copy of the contract with your contractor and any other documents your lender may require (for instance, a copy of your original title policy, copies of bonds, a construction cost breakdown). Inform the other lenders of your selection of a lender.

DO:
- Arrange construction financing early.
- Submit loan applications to several lenders. Shop intelligently and vigorously for the best terms. You can save a bundle!
- Ask prospective lenders for a "Construction Loan Checklist."
- Organize all required forms neatly; make copies of every form you submit.
- Tell the truth on your financial statement. Don't exaggerate, but don't be bashful, either.
- Scrutinize the checklist for reasons given above why your loan application could be denied. See the stumbling blocks?
- Consult an architect, attorney or realtor before accepting a loan.
- Insist that any portion of your own money deposited with the lender as part of the loan package draws interest from the lender until your portion of the money is disbursed to the contractor.

DO NOT:
- Agree to pay "dutch interest." Ever!

TIMELY FOOTNOTE:

We have mentioned above the typical, the traditional loans, which were standard until recently. During the last two years, however, financing has changed: fund sources dried up, interest rates went beyond the absurd to the impossible. There is now—and there will remain—stiff competition for construction and real estate funds, both from industry and from government, especially for long-term loans. This means the lenders, the builders and contractors and you, the consumer, we all have to realize, it's goodbye to easy, ten-percent, 30-year mortgages and hello to new types of loans. For instance:
- Seller or contractor will "carry paper". (Second or third mortgages/trust deeds, with balloon payments)
- Variable rate. (Long-term but rate may increase or decrease every year)
- Graduated payment. (Lower monthly payments for five years, then higher than normal payments)
- Rollover. (Interest rate is negotiated every five years)
- Shared appreciation. (Lender or contractor receives ¼ to ½ of the building's appreciation in exchange for 3% to 6% lower interest rate)
- Growing equity mortgage/trust deed. (Monthly payments will raise 3% each year; the increases are applied to the principal of the loan)
- No interest. (Are they kidding?) (Large down payment plus so many equal payments, normally 60 or 72; loan is paid in five or six years. Bingo.)

With such a variation of new types and terms of loans, the borrower is entitled to look confused. Once more: put all those pretty numbers neatly on paper and compare and shop and shop and shop. Also: get expert advice—from an accountant or realtor.

B-2 HOW TO FIND A GOOD CONTRACTOR

A good contractor is as important as a suitable building site, watertight drawings, specifications and reasonable financing There is no substitute for a good contractor, even for small projects. By "contractor" we mean general contractor, specialty contractor or subcontractor whichever fits your needs.

A good contractor:
*   is experienced in the type of work under consideration
*   has the time, personnel, insurance and license to take on the project
*   is financially stable and can furnish references
*   is congenial (better yet, enthusiastic!)
*   can do the work at a reasonable price

If, in addition, he:
*   submits a low bid (not necessarily the lowest)
*   can make a few suggestions in regard to your drawings and specifications and cost savings
*   has a happy homelife and is a member of the Optimist Club so much the better!

Nothing is gained and a contractor can become a nuisance if he:
*   is your friend, neighbor or brother-in-law
*   has never done similar work but would like to try
*   needs work desperately and asks for your mercy
*   promises to complete the building in 40 days or less
*   says he has coffee every morning with the building inspectors.

Him you don't need!

How, then, to find the "good contractor" for the small to medium-sized project: residences, commercial structures, apartments, churches?

► Call the local Building Contractors Association, describe your project and ask for three or four names. Or ask for a roster; these trade associations have lists of every type of general contractor and subcontractor, divided by categories such as remodeling, shell construction, demolition, etc.

► Your architect should come up with two or three names. If you don't have an architect, phone several and ask them to give you contractors' names. Ask also estimators in lumberyards, building inspectors, loan officers.

► Drive around and look for structures under construction that are similar to yours. Get the contractor's name from signs, trucks, the guard, workmen, neighbors. Same holds true for finding subcontractors: tour the neighborhood.

► If you know any general contractor who does large projects, ask him or his estimator to refer to you a contractor who does smaller jobs.

► Ask friends, colleagues and employers who have built recently.

► Scan the building section of your local newspaper, especially the display ads in the weekend edition. Remodeling contractors, for instance, prefer to advertise there. After having read Chapter A-7, "Contractor Provides Free Drawings," you probably will be careful about advertisements that offer "free design, free permits, free whatever." But among reputable remodeling contractors you could find the fellow with the expertise, the crew and the staying power for your project.

► Picking names from **classified ads?** Last resort. It's a frustrating experience!

► You could advertise yourself in the classified ad section, under "Contractors and Builders." Yes, we know that is a Services Offered section, but a lot of contractors will read it if only to see whether or not their ad appeared there correctly. Or what their competition has to say. So all of a sudden, here comes your ad:

> GENERAL BUILDING CONTRACTOR WANTED
> TO SUBMIT BID TO BUILD 2000 SQ. FT.
> RESIDENCE WITH GARAGE. MUST HAVE
> REFERENCES. YOUR NAME AND TELEPHONE

Our definition of a good contractor for large projects is the same as for smaller ones, only the method of finding him is somewhat different: You should call the General Contractors Association for names and then depend heavily on referrals from the business community.

When should you start looking for a contractor? Collect names as soon as you have decided to build. Sure you have weeks and months until you have to select a contractor, but start looking early. There are good reasons: One of the contractors could be gainfully involved with preliminary cost estimating during the planning phase. Also, if you wait until you need the bids, it always seems to take longer to eventually find three or four contractors willing to bid.

All right, you have a list of good prospective contractors, say seven. You have phoned them. Five were interested in bidding or at least in further discussion, one never returned your call (a telltale; to hell with him!) and the last guy's number is not in service anymore and there is no new number.

What next?

Personally meet each one at one of his building projects or in his office, even if it is a home office. Better yet, meet them (at different times of course), at **your** jobsite, so they can look over the place, especially if it concerns a remodeling. Ask these questions:

• Is he willing to bid on the project?
• How many sets of drawings and specifications does he need? How much time for bidding? Where do you send the bidding documents, or can he pick them up? Tell him there is a refundable deposit, equal to the printing cost, for each set.
• Can he furnish addresses of projects similar to yours, including names and phone

numbers of owners? Other references are helpful: architects, his banks, building associations.

When interviewing contractors, there will be at least one who will try to talk you into negotiating a contract with him rather than going through the rigmarole of competitive bidding. Don't commit yourself to a negotiated contract for any of the reasons the man may give you: free design, financial help, deferred payments, etc. There are occasions where a negotiated contract is appropriate; we shall talk about that later. In most cases, it is not advisable to forego competitive bidding and just negotiate a contract with one contractor.

DO:
- Start early to compile a list of good contractors.
- Meet each contractor in person; get references and pay attention to your gut feelings. Insist that he visit your building site.
- Give preference to "competitive bidding" rather than to "Negotiating a contract." Why? Chapter B-3 explains.

B-3 COMPETITIVE BIDDING OR NEGOTIATING A CONTRACT

Act I: One man we know wanted to add a large living room to his 2-story residence, convert the existing living room into a formal dining area and add a family room downstairs. A former schoolmate of his was a building contractor—reliable, capable and successful in remodeling work. He proposed negotiating a contract for the remodeling work, including free design, permit drawings and obtaining the permits. The estimated cost was between $24,000 and $26,000, depending on the selection of material by the owner. It sounded reasonable.

Act II: Out of curiosity, the owner asked what we thought. Our reply indicated we thought he was in good hands with his contractor, but we wondered about two aspects:
1. Would his contractor friend show him more than one way to design the addition?
2. Was it possible that another contractor could do the same work for $1,000 or $2,000 less, and with that difference the owner could buy new carpets or drapes that were not part of the remodeling contract?

The answers:
1. No. His contractor friend had made one sketch containing the number of square feet they had agreed on. As to the architectural pizzazz, *the contractor's expression,* he said, "Nothing to it"; they would just match the existing building—no Frank Lloyd Wright needed here!
2. Probably not. Both considered the price range as being in the ball park. Would the contractor, as a friend, take advantage of the owner? Hardly. The contractor would show him all the material invoices, time cards, subcontracts, expenses and profit.

We had one more question: Even if he wanted this particular contractor to have the job, could it be that by putting it out for bid the contractor would sense the competition and lower his price?

The answer from the owner: Since the contractor did the design and drawings free, he would not allow the use of them by other contractors.

"Not even to bid on them?" we asked.

"No."

"We thought the design was free."

Act III: At this point we proposed designing at least two, maybe three solutions and discussing with the owner the advantages, disadvantages and costs of each solution, then producing working drawings and specifications. The owner would pay for plancheck, permits and put the documents out for competitive bidding. Our proposal was accepted.

Since all the contractors who were willing to bid were good contractors we suggested that the owner give the job to the lowest bidder. (We would not knowingly put any contractor through the process of bidding if we didn't think he had a genuine chance of being awarded the contract.)

Based on our experience, we were convinced that the sum of the best bid plus all permits plus our architectural fee would be lower than his friend's proposal. The owner was happy with the designs we produced. We settled on one he considered very functional and unique, with more square footage than he ever thought he would get. Here are the bid results:

Contractor A	$31,600
Contractor B (his friend)	25,800
Contractor C	23,200
Contractor D	22,560

The bottom line cost to the owner at this stage:

Contract with "D"	$22,560
Permits	145
Architect—for design, contract documents, bidding, selecting the contractor	2,057
TOTAL	$24,762
The friend's bid	25,800
To the client, so far: better design and net saving	$ 1,038

Comments:

In general, competitive bidding brings costs down at least 10 percent. In this somewhat simplified illustration, the savings was higher.

The friend's bid, including permit fee	$25,800
Contractor D plus permit ($22,560 + $145)	22,705
Gross savings	$ 3,095

Savings related to friends's bid:

$$\frac{\$3,095 \times 100}{\$25,800} = 11.9 \text{ percent.}$$

If you factor in the architect's work, the net savings to the owner still is:

$$\frac{(\$3,095 - \$2,057) \times 100}{\$25,800} = 4 \text{ percent,}$$

which represents the net savings of $1,038.

The bottom line: the architect's work came free in addition to the net savings and much better design and construction.

The message, in capital letters: THE ONE MOST EFFECTIVE DEVICE FOR FINDING

THE LOWEST POSSIBLE COST FOR ANY TYPE OF WORK IS COMPETITIVE BIDDING!
There are occasions where a negotiated contract is appropriate, especially the "cost-of-the-work-plus-fee" arrangement (see Chapter B-11). But in seven out of 10 cases, it is advisable to stick with competitive bidding that will result in a lump sum contract.

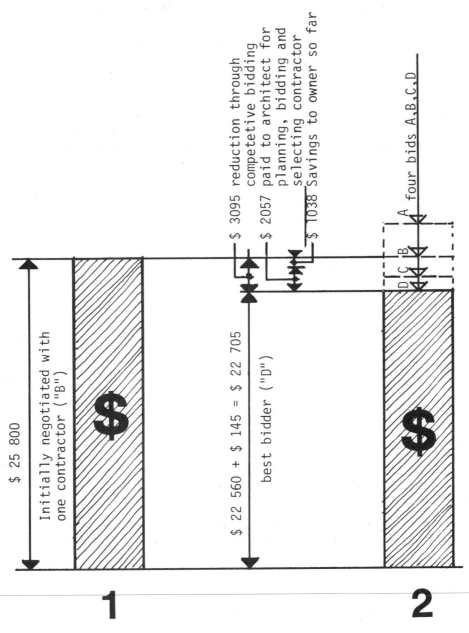

1

SINGLE BID
Negotiated with
one contractor

2

COMPETETIVE BIDDING
the most effective
devise to arrive at
the best price.

How to obtain meaningful bids?
Four points to consider.
Coming up.

B-4 TIME TO GET BIDS

There are two well-known methods to obtain a proposal for your project:

1. Direct selection of a single contractor, resulting in a negotiated contract with a lump sum.
2. Competitive bidding between several contractors, resulting also in a lump sum bid.

There is a third way, not so common, a clever combination of these two methods. It is called the "cost-of-the-work-plus-fee" arrangement: owner and architect select a general contractor, negotiate a fee for his work as a coordinator and the general contractor in turn solicits competitive bids on the subcontracts. Together with owner and architect, he selects the subcontractors.

Every method has advantages and disadvantages, which will be discussed in following chapters. In probably seven out of 10 construction cases, method 2, competitive bidding, is used. Therefore, this chapter deals with competitive bidding, which is the most effective method to find the lowest possible cost for any type of work. Some of the principles discussed also pertain to the task of negotiating a contract, for instance, the notes on the best time to get a proposal and how much time to allow for bidding.

All principles pertain also to the third method, the cost-plus-fee contract. However, this type of agreement is discussed in a separate chapter, B-11, not only because it merits special consideration but also because it entails some specific arrangements.

So, let's talk in general about bidding between several contractors, competitive bidding. There are four questions to consider:

What is needed to get competitive bids?
When is the best time for bidding?
How many bids should you get?
How much time do contractors need to prepare the bids?

Your architect should know the answers, but in case you are curious or decide to go it alone, here they are.

What is needed

Competitive bidding means shopping intelligently for the best proposal. Let's compare with something most of us are familiar with—shopping for a car. You want to buy from a dealer (i.e., contractor) who is reputable, gives certain guarantees and is still around when something goes wrong with the car (i.e., building). Beyond that, how do you get the best price? Describe the car you want to two or three dealers, ask for a quotation and compare prices. When comparing dealers' (contractors') proposals, you want to be sure that each bid is based on the same product (building): same model, material, color, horsepower, amenities, taxes, delivery, guarantees.

An oversimplified exercise, we admit, yet there are at least three points to make:

a. To find the lowest price, you have to tell several contractors what you want, ask for a quotation and compare.

b. To compare intelligently, all quotations must be based on the same requirements (drawings and specifications).

c. Considering past performance, reputation and guarantees, and to avoid extra costs, misunderstandings and other disappointments, the **lowest bidder** is not necessarily the **best bidder.**

To get meaningful competitive bids, the contractor needs:

► instructions to bidders (bid date, time, place)

► precise, watertight, clear drawings

► specifications, with general and special conditions

► sample of owner-contractor agreement

► proposal forms he should use

During the bidding, the owner or his architect must be available to answer contractors' inquiries and issue addenda to all contractors if necessary.

All bids shall:

► be submitted on the prescribed proposal forms, dated and signed

► indicate the type of contracting firm—single proprietorship, partnership or corporation

► be good for at least 60 days

► acknowledge that addenda, if issued, were received

► state completion time in calendar days and that the contractor has visited the site

► list all subcontractors

► be accompanied by the bidding documents (drawings, specifications, sample contract) which should be returned to the owner.

Best time to get bids

Most projects go out for bid at a time governed solely by the owner's requirements. He wants the building in June, therefore construction should start the preceding November; since four weeks each are scheduled for obtaining bids and selecting the contractor, the bid documents go out for bid around September 1.

In bidding on construction projects, there is one characteristic similar to the stock market: most people buy stocks when everybody else is buying. In building, most people plan to build when everybody else is building. Our advice, in capital letters:

BUILD WHEN THE OTHER GUY ISN'T!!

This bland counsel is usually not very successful because to build when nobody else does, during a recession for instance, takes guts. But we continue to advise remodeling homeowners, church committee members and corporate planners to get bids and build when others do not. More on this in Chapter C-1, Best Time to Build.

How many bids to get

From three, four or five general contractors; each general contractor from subcontractors: at least two from each trade or vendor. To be able to compare proposals, you need, obviously, at least two bids. Three or four are better. If you have only two subcontractors' bids on carpentry, it is difficult to get a feeling for the realistic cost range unless the proposals are close to each other. If they are far apart, the low bid could be dangerously low: the fellow might have forgotten to include page 2 of the lumber list. Don't trust a bid that seems to be too low.

To obtain enough subcontractors' bids, ask the general contractors who are bidding on your project if they wish one set of drawings, specifications and addenda to be placed in the contractors' plan room in your town.

A question often asked is should you tell a contractor who else is invited to bid?

Yes. Good contractors often do not bid if they don't know who their competitors are.

How much time to allow for bidding

Depends on the construction activity in your area; if it is hectic, allow one or two more weeks than normal. We have found that if you allow too much time, even with a specific reason, most contractors will not huddle over your drawings until the last week before the bid date.

One particular reason for extending the bidding time is when a few other projects of your kind and size are due for a proposal on the same date. Check your local Construction Reporter for bid dates. If you change the bid date, inform all general contractors immediately.

These are the average bidding times we allow general contractors:

Type of building	Weeks	Number of bidding documents to each general contractor
New residences, residential remodelings, simple industrial buildings	3	3 to 4
Office buildings, churches, apartments, motels, hotels	3 to 3	6 to 10
Schools, large industrial buildings, high rise hotels, hospitals	4 to 6	10 to 24

Experienced contractors, once invited to bid, will still try to find out how serious the owner is to build, if he can obtain financing, who else is bidding and if he, the contractor, has a genuine chance of landing the contract or is just being used as a stalking horse. Most architects observe the sensible rule to not use a contractor in vain, not simply out of a sense of fairness, but out of common business sense. The architect would never get the contractor who was misled to bid on one of his projects again.

On small owner/builder jobs, the bids for certain trades or equipment can be had in hours; carpeting, for instance, or kitchen built-ins, concrete yardage, flooring, windows, etc. Other trades take days or weeks—lumber, carpentry, plumbing. The layperson seldom appreciates how much time and effort it takes to put together a complete, responsible proposal, with bids from at least two subcontractors or vendors for each trade or equipment.

Under time pressure, general contractors get a proposal from only one subcontractor in many of the trades, a problem the owner should be aware of and be prepared to deal with effectively. Chapter B-13, Interchanging Subcontractors, tells you how.

DO:

- Have precise, but not overly detailed bidding documents (instructions, drawings, specifications, sample contract, proposal form).
- Choose the best time for bidding and building—a recession, for instance.
- Issue enough sets of bidding documents to contractors.
- Consider placing one set of bidding documents in the contractors' plan room in your city.
- Allow reasonable bidding time; extend for good cause.
- Issue addenda if necessary.

Competitive bidding brings the cost down. Not only in the construction trades. Get competitive bids on— almost everything. See B-5.

B-5 GET BIDS ON—ALMOST EVERYTHING

The Story:

To get a construction loan, our bank insisted that we provide "Fund Control," a special loan disbursement service, described in Chapter C-4. Get either Company A or Company B, our secretary was advised. These were the two firms almost exclusively recommended by this particular bank. Not that the loan officer insisted that we use either one of the two; he did not. By law the lender cannot insist on using a particular fund control firm, insurance carrier or title company. It was just routine advice: "Here are two names, we use them all the time." Since fund control fees are paid by the borrower—that is, us—we have the right to select the fund service firm. We knew four firms in town who provided the service; a realtor friend gave us a fifth one. Our secretary phoned all five companies and shopped for a quotation:

"We need fund control for a construction loan of $60,000. Would you give us your rate, please."

It took her about 20 minutes to make the calls to the five firms, each time asking for the Fund Control Department. Here are the bid results:

Company A $300.00
Company B 300.00
Company C 265.00
Company D 225.00
Company E 175.00

The difference between the first two (recommended by the lender) and the low priced firm, which we used, was $125. One hundred twenty-five dollars saved by competitive bidding, so to speak, with five phone calls, in 20 minutes. Where can you save money faster?

When carefully done, more often than not with professional help, competitive bidding can be applied to almost any activity in the building process: bank loans, soil reports, topographic surveys, land development schematics, attorneys, accountants, architects, consultants, fund control, general contracts, subcontracts, cost-plus-fee contracts, equipment rental, maintenance agreements, realtors.

The message:

For almost any material, item, labor or service there is the lowest possible price, the best bidder. Shop for it intelligently by "competitive bidding." In the previous chapter we

explained the procedure:
- Collect names of reputable firms.
- Contact them, specify your needs, ask for a quotation.
- Compare the prices; be sure they are based on the same scope of work and provide identical service.
- Give the job to the best bidder.

A word of caution:

In construction the task of shopping for competitive bids ranges from easy to most difficult. It is easy, for instance, to shop around for a built-in trash compactor: Just get the model number, phone three supply houses and be sure they all include or exclude in their quotation the same variables (model number, taxes, color, guarantees, delivery, etc.). To shop for professional services is the most difficult and we suggest you give high priority to reputation, recommendations by others and a certain personal gut feeling and not price alone. This is especially true when selecting engineers, architects, accountants, investment advisers, realtors and others.

By now we hope you have posted at least two signs next to your mirror:

Sign No. 1: DECISIONS—DECISIONS—DECISIONS

Sign No. 2: The one most effective device to find the lowest cost of any type of work, material or service is competitive bidding.

GET BIDS ON—ALMOST EVERYTHING

In order to evaluate bids, they must
be based on equal performance.
Sounds easy, but isn't.

B-6 SUBSTITUTIONS, EQUALS AND ERSATZ

In Chapter A-8, Who Needs Specifications, we told you about the kind elderly lady who had visions, not matched by the tin can vision of her contractor, of a new bathroom, right out of House Beautiful. It came to that traumatic case of ersatz because the flimsy drawings did not define the materials to be used. This happened, as you might recall, in the "planning and permits" phase. Similar pitfalls abound in the bidding phase. The contractor sees a certain item or room described on the drawings and specifications, but has in mind to eventually provide something different; sometimes a little different, sometimes dramatically so. This happens for a number of reasons:

a. The contractor is not familiar with one product or has qualms about it and has never installed it, but will provide something similar, which he considers as good, or in construction terms, "equal." Let's say you specified Pozzi-Ginori-Milano plumbing fixtures, but the contractor always uses American Standard fixtures (or vice versa). He knows them best, trusts them. He wouldn't use anything else and swears by their quality, guarantees and availability of spare parts. The cost is not the same, however, so the bids are not based on the same work; therefore, his bid is difficult to compare to the contractor who followed the specifications and bid Pozzi-Ginori fixtures.

b. The contractor is told by his wholesale house that it cannot get Pozzi, there is a strike in Milano or the wholesaler does not handle the brand. The plumbing contractor really tried, but the general contractor pushes him to get his act together, install the plumbing and get out. So American Standard is regarded as an equal and will be installed.

Right or wrong? In both cases the substituting contractor probably stands with one foot on legal ground. Most architects' specifications state: "Items specified on drawings and specifications represent the type and quality required. Contractors may substitute items considered equal in their bids when approved by the architect."

The last part of that sentence indicates that the other leg of the substituting contractor is on shaky ground if the substituted item is not approved by the architect as being equal. If it is approved by the architect—or engineer or owner/builder—but not considered equal in quality, a monetary adjustment must be made by the contractor and the bid must reflect that adjustment. The surest method is to require contractors to bid what's on the drawings and specifications so the owner looking at four bids, really compares apples to apples to apples to apples. On a separate sheet attached to the bid, substitutions or alternate methods of construction should be tabulated showing an amount to be added to or deducted from the "base bid" in case the substitution is accepted.

c. Substitutions can be invited by the architect! In order to become aware of any way the building cost can be lowered, we place in the "Instructions to the Bidders" the following note:

Substitutions

Bidders are encouraged to suggest modifications of methods or materials that would reduce the cost of construction without changing substantially the layout, features, quality level or appearance of the building. Such modifications, and the amount to be saved by the owner, should be listed on a separate sheet and attached to the bid proposal.

Two more reasons for substitutions:

d. **During bidding** the owner or architect, or both, changes his mind; first he wanted Pozzi-Ginori fixtures, then he decided to use American Standard. In this case the architect should issue an addendum specifying the switch from one brand to another, and send it to all bidding contractors.

e. **During the construction period** the contractor finds out that he truly cannot get the specified fixtures, although he did bid the brand; therefore, architect and owner settle for the other make. Since this substitution takes place during construction and therefore affects only one plumbing contractor, the architect shall prepare a written change order describing the change and noting any adjustment in cost and construction time. Owner, architect and contractor will sign the change order.

Hold it! There is one more substitution case:

f. Specifications call for a **5'-6" cast iron tub** and the plumbing contractor installs a **5'-0" steel tub** with a painted-over chipped spot. He assumes the owner will never know the difference and the architect will never come for inspection.

This is a clear case of cheap ersatz. The architect or owner shall call the contractor, even if it is Sunday morning, and demand replacement by Tuesday night. In addition, the owner should check every stick and stone this contractor has installed so far to make sure everything complies with drawings and specifications.

DO:
- When receiving the bids, verify whether substitutions are assumed or listed by contractors.
- By a note, encourage contractors to make suggestions for alternate methods of installation, materials, etc.
- When drawings or specifications are changed during **bidding** have an **addendum** issued.
- Have a **change order** issued when the change occurs during **construction.**
- Understand the value of regular inspections by a knowledgeable person.

DO NOT:
- Tolerate ersatz. Call the contractor immediately. Get tough!

WERNER R. HASHAGEN & ASSOC.
ARCHITECTS
7480 La Jolla Blvd.
La Jolla, California 92037

Tel. 459-0122

TO   General Contractors

LETTER OF TRANSMITTAL

DATE 4/29/81	JOB NO. Gillispie
ATTENTION	all bidders
RE	Instructions to bidders
	Addendum No. 2

ADDENDUM No. 2 (Page 1 of 2)

Special Conditions: Owner will pay directly for: Building Permit, Water meter, electric meter and connection from building to SDG&E Co. Line, sewer connection, if any, Engineering sidewalk permit and bond.
Subcontractors will pay permits for their respective trades: electrical, plumbing, heating, venting etc.

Sheet A-1   Gasmeter will be located in planting area at N-E corner of building. S.D.G.&E. Co. will furnish and pay for gasmeter and line to meter, including trenching, backfilling and repaving.

Furnish and install approx. 18 linear feet of handrail and attach to existing wood fence. Handrail from planed 2x4, kiln-dried D.F. or similar, with 5 standard metal brackets. Paint or stain handrail 2 coats.

Sheet A-3   In exterior concrete slabs on ground (walks, patio, entry slab): install weakened plane joints, approx. 6 feet on center, 1/8" x 1 inch deep.
Install 4 inches of coarse sand under all concrete flatwork.
Install 6x6 - 10/10 W.W.M. in two patio slabs only, at midslab.
Typical wall footings can be 12" wide in lieu of 15".
Typical pad footings shall be 24" square in lieu of 18".

Sheet A-4   Doors No. 6 and 7: to be 3' - 0" wide.
Doors No. 4 and 10: to be 20-minute doors. Bid T.M. Cobb (or equal) Mineral core, 1 hr. natural birch, B-Label, 1 3/4 thick; with Door No. 13 not a rated door anymore. Metal astragal, in steel frame.

Both restrooms must be accessible to the handicapped persons:
Install grab bars in two toilet stalls total, in one of each restroom.
Total of 2 stalls must be 3'-6" wide each.

Delete: notes regarding one-hour floor.
5/8" plywood is sufficient plus floor covering as specified.

Sheet A-5   On soffits over niches, windowseats: use medium spray in lieu of acoustical plaster.

On classroom ceilings: 12x12x1/2 T&G acoustical tiles
Armstrong No. 715, fissured monotone
in lieu of 24x24 Colortone (not available).
Glue: Webtex all-purpose acoustical adhesive.
Source: Maloney Specialties, 275-0660.

Sheet A-6   Underfloor fiberglass insulation shall be held in place with laced wires.
Topchord of trusses to be from 2x6.

Sample of an addendum

Aside from substitutions, most bids contain cost items that are not clearcut: work or services which later, during construction, can cost less or more. If they cost less, do you get money back? If they cost more, who pays the difference? Why, in the first place, is the cost ambiguous? The next chapter on "Allowances" explains.

B-7 ALLOWANCES

In the planning stage the owner will not make, for various reasons, all the decisions and give the architect all the input necessary for complete drawings and specifications. Even the architect and Building Department plan checkers will defer decisions. That means there will be blank spots when drawings and specifications are sent out for bid, and later, for actual construction.

In order to satisfy the lender's requirement that the owner submit a complete construction cost breakdown and also to assure the owner of a reliable maximum construction cost, amounts for the undecided "blank" items may be estimated and entered as "allowances." For instance:

- Soil conditions may not be entirely known until trenches are excavated.
- Finish hardware or kitchen built-ins may not be selected by the owner yet.
- Built-in drapery rods: The owner cannot decide whether he wants them or not.
- Carpeting will be selected by the owner, probably even under a separate contract; but the item should be mentioned in the specifications so an estimated cost can be entered into the cost breakdown sheet for loan purposes.
- Certain construction items are **almost always** entered as allowances: contingencies, finish hardware, bathroom accessories, light fixtures, kitchen built-ins, wallpapering, cost of utilities during construction.
- And these materials and equipment are **often** listed in the form of allowances: wood trusses, plumbing fixtures and fittings, resilient flooring and bases, ceramic tiles, sprinklers, intercoms, antennas.

Experience tells how much should be allowed for these items; the actual cost later on may be less or more and an adjustment will be made before the final payment goes to the contractor. Let's look at finish hardware as an example. Often there is not enough time to furnish a complete hardware schedule for bidding. The owner reserves the right to visit showrooms and then make his selections, especially in residential work. Whatever the reason, an allowance should be estimated and inserted in the specifications and the cost breakdown sheet. A typical finish hardware allowance for $105,000 residence construction cost could be 1 percent or $1,050. The architect or an experienced hardware supply house should advise on this. The allowance would encompass, for instance, all hardware for exterior and interior doors, including pulls, locks and hinges for cabinets and pullmans; and often towel bars, paper holders and other bathroom accessories.

Another typical allowance is for "contingencies" 3 percent for new work, 8 to 10 percent for remodeling. When allowances are specified, the text must make clear a number of aspects:

- Is the allowance for material, delivery and taxes only? In most cases handling and installation are part of the sum of the total contract, not of the specific allowance.
- If actual costs later are more or less, an adjustment shall be made by written change order; the owner may have to pay additional monies or get a refund in the form of a check or by a deduction from the last payment to the contractor.

Under the allowance method, the owner will select certain materials during the construction phase. The contractor will tell the owner how much time (or how little!) he has to make those selections. Our advice: Select materials covered by an allowance as soon as possible after construction starts. Contractors get irritated if the tilesetter is scheduled for Monday morning and by Friday night the owner still hasn't decided whether he fancies 4" X 4" tiles, 6" X 6" or 3" X 6"; Spanish or Dutch; laid in mortar or glued on; avocado green, seagreen or bluegreen.

DO:

- Make use of allowances for items, materials or work not specified in detail by the time the drawings and specifications go out for bid or even for construction.
- Insist that monies allocated in allowances be accounted for by written change orders from the owner or architect and by itemized vouchers.
- Unused portions of allowances shall be returned to the owner before the final payment is made to the contractor either by check or by a deduction from the last payment to the contractor.

"All clear now on bidding documents?"
"Sure! Instructions to bidders, proposal form, drawings, specifications, sample of owner-contractor agreement."
"Good show! — Any more questions?"
"Yeah. When we know how much this whole affair is going to cost?"
"Monday! Next Monday is the bid date."

Shirts and shoes, please: . . .

B-8 BID OPENING

Let's talk now about bidding and instructions to bidders. The instructions should be brief, on one or two sheets of 8½ x 11-inch paper and give the following information:

1. How many free sets of drawings and specifications each invited bidder will receive, against a refundable deposit. The deposit will be about equal to the printing cost and refunded when the sets are returned in good condition on or before the bid date. These sets can be used later by the successful contractor during construction. Extra sets will be issued at printing cost, which the contractor will pay. They do not have to be returned.

2. Bid date, time and place; list of bidding documents.

3. Whether bid opening is "public" (all bids are read aloud, with bidders present) or "private" (owner opens bids with architect, at their convenience).

4. Whether or not bonds will be required. See also B-15.

5. In order to not miss any possibility of reducing construction costs, we include this note:

   Bidders are encouraged to suggest modifications of methods or materials that would reduce the cost of construction without changing substantially the layout, features, quality level or appearance of the building. Such modifications, and the amount to be saved by the owner, should be listed on a separate sheet and attached to the bid proposal.

6. Bid openings for government work are formal and follow punctilious rules. For private projects, where the owner spends his own money, bid submittals are less formal and don't have to last longer than 20 minutes. The steps that lead to the bid "opening" are as follows:

   • From a list of seven general contractors, five indicated they were interested in bidding. The five contractors picked up three sets each of drawings and specifications at the architect's office. The architect spent about 15 minutes with each bidder to briefly explain the project. Each contractor left a check of $24 as a deposit on the drawings and specifications.

- Three weeks were allocated for bidding. During this time the architect sent to all five contractors one addendum with modifications the owner wanted and some architect's clarifications of the drawings. Contractors also asked that one set of drawings and specifications be placed in the city's plan room so that any interested subcontractor could submit a bid to any of the listed general contractors.

- Bid date was a Monday, 6 p.m. deadline, at the architect's office. If the owner wishes, it could take place at his office or his home. Monday was chosen because it gave the contractors an additional weekend; 6 p.m. gave them another full day to assemble the proposal from last-minute bids and allowed them to avoid rush hour traffic. None of the contractors had asked for an extension of bidding time.

- On Friday, before the bid-Monday, the architect's secretary phoned bidders to inquire if they had received the addendum and if they had any questions at that time.
  "No more questions, gracias."
  "We will see you on Monday at six then?"
  "Si."
  The phone call served not only to make sure all had received the addendum, but also to let the architect know if he could hope to get five proposals. One contractor did say he probably would not bid.

- Public bid opening was at 6:10 p.m. with three contractors present; one had delivered his bid Monday morning in a sealed envelope. The fifth contractor did not bid; he also did not return the bid documents. Deposit checks were returned to three contractors; the fourth check was mailed out, the fifth one was later cashed by the owner since drawings and specifications were never returned. Before opening the bid envelopes the architect reminded the three contractors that all bids had to be signed and the addendum acknowledged in writing. Then he read the bids and alternates aloud; the secretary entered all numbers in a bid tabulation. Here is a sample of a bid tabulation form:

| Contractor | Addenda rev'd. | Base bid | Add or deduct | | | Base bid & alternate proposal combination | | |
			Alt. #1	Alt. #2	Contr. Suggestions	B + 1	B + 2	B + 1&2

One of the four bids was clearly the low bid, not only as a single amount base bid, but also as the sum of any possible combination of base bid and alternates.

- While the contractors drank the architect's coffee, the secretary made copies of the bid tabulation and gave one to the owner and each contractor.
  6:32 p.m.—Owner thanked the contractors for their effort and said he would discuss the bids with the architect soon and let all bidders know of his decision within a week or so.

6:41 p.m.—Everybody faded away. That's the advantage of the late afternoon
bid time: Nobody hangs around too long; all are tired or hungry. Or just plain thirsty!

```
P R O P O S A L

To:



This contractor has carefully examined the construction documents:
     Drawings and Specifications Dated 8/29/81
     Addenda No. _____

and has visited the site and examined all conditions affecting the work.

I/we propose to furnish labor, materials, tools, equipment, insurance, taxes and
other related items and services required to properly execute and complete the
construction in conformance with construction documents prepared by Werner R.
Hashagen & Associates, for a
LUMP SUM BID of $_____  (_____
                 _____Dollars)

Alternate:
If the owner prefers a cost-plus-fee contract, the
FIXED PROFESSIONAL FEE to the General Contractor will be
              $_____  (_____
                 _____Dollars)
with the contract based on the A.I.A. cost-plus-fee form A111-1978.

Time:  I/we propose and agree to substantially complete the building and
       related work within _____(   ) calendar days.  We can start
       construction on _____.

Insurance:  Public liability insurance will be placed with
            _____

            Workmans Compensation Board insurance with
            _____

Bonds:  If directed by the owner, I/we will deliver to the owner, within ten (10)
        calendar days after signing the contract, a payment and performance
        bond, in a form and with a surety company acceptable to the owner.  The
        cost of such bonds will be paid by the owner.

This proposal is good until _____, 1981.

CONTRACTOR: _____

            Individual  /   DBA  /   Partnership  /  Corporation

            _____
            Signature, Title
                                         Seal

Address: _____
Telephone: _____   Date: _____
Attached:  List of Proposed Subcontractors          Proposal
                                                     1 of 2
```

Typical Contractor's Proposal Form
(Furnished by the architect to all bidding general contractors.)
The second page of this proposal is a list of subcontractors
the general contractor would use if he were selected as the best bidder.

Later on the owner will read the bid tabulation a number of times with his architect,
his accountant, his spouse. Strange. Some
of the numbers look weird, out-of-line.

Contractors' bids have five ingredients:
labor, material, expenses, overhead and profit.
If several contractors in the same town, at the
same time, under the same tax rules, bid on the
same brick, mortar and sticks, why then, do we
get these large spreads? . . .

B-9 THESE UNBELIEVABLE SPREADS IN BIDS

Shop around for a car tire, let's say for a Volkswagen Squareback. In order to compare apples to apples, you jot down a sort of specification of the tire you want:

one blackwall tire; 165-15R; steel-belted; installed, including balancing; weight and stem valve; including all federal and state taxes; with (or without) trade-in

Then you shop. The price quotations from four tire dealers were:

Dealer A $56.45
Dealer B 60.28
Dealer C 60.52
Dealer D 65.27

This competitive bidding for a car tire by phone took 25 minutes and saved the owner money because he found the best bidder, Dealer A. Could he have known about A if he had not asked for several quotations and then compared them? Only by obtaining and comparing bids can you find the best deal.

Why do we talk about car tires? To show that in most competitive markets like tires, furniture, cars and stereos, prices differ, but not really dramatically. Look at the tire prices again. If you add all four quotations and then divide the sum ($242.52) by four, you get $60.63, which is the average quotation. Can you bear high school math for a moment? The highest quotation of $65.27 is about 7.6 percent above the average of $60.63; the lowest, $56.45, is about 6.8 percent below the average. Add up these two percentages and you'll find that the total spread between the high and the low is about 15 percent (or, to be exact, 6.6 plus 6.8 equals 14.4 percent).

If you don't count in distress sales, bankruptcies and loss-leader giveaways, prices for most articles are really not that far apart.

Until you ask for bids in construction! Here quotations have a wide spread, often unbelievably so—in the same town, at the same time, for the same work, based on the same drawings and specifications. Take a look at bids for one of the smallest projects we've had to bid.

The work to be done:

> Enlarge an existing garden wall: lay 255-4 x 8 x 16 regular concrete blocks, with a total of 100 lineal feet of 6'' ladder mesh. Top the wall with 40 cap blocks to match. Contractor to inspect wall before bidding. Painting not included.

And here are the envelopes!

Contractor A	....	$970
Contractor B	....	620
Contractor C	....	425

All had the same license, could do the work within the next two weeks, would leave the place broom clean, and would probably buy the blocks and other materials from the same supplier. But the bids were far apart. Let's calculate the average bid:

$$\frac{(\$970 + \$620 + \$425)}{3} = \$671.$$

The highest quotation of $970 is about 45 percent above the average of $671. The lowest bid, $425, is 37 percent below the average. The total spread between high and low is unusually high, about 82 percent.

We used to think that lousy drawings and specifications were the reasons for wide spreads in bids. It is true that incomplete bidding documents produce wild proposals. There are several reasons:

- The skinflint contractor bids strictly what is shown on the drawings and in specifications, if there are any! He will bid on the lowest quality level consistent with the incomplete drawings.

- The more conscientious contractor has a problem. He has to guess what should be shown on the bidding documents, but isn't, what could be accepted by the owner and what would satisfy the building codes and inspectors. He will also study the owner's financial standing to see if the owner will eventually pay for extras.

- The good contractor has to add a hefty amount for contingencies; his bid will be safe but out-of-sight!

- Or worst: The good contractor doesn't bid at all or will try to negotiate a contract, hoping to fill in the missing parts when he can sit down with the owner.

However, skimpy drawings are not the sole reason for wild numbers in the bid lineup. Look at the proposals received by the Navy, the telephone company or the school boards. The spread in bids is difficult to explain since these agencies have precise bidding documents, with time-tested and carefully delineated standard details. Here is a recent bid result of such a government project:

General Contractor A	....	$33,989
General Contractor B	....	42,000
General Contractor C	....	57,987
General Contractor D	....	72,375
General Contractor E	....	74,454

The spread between high and low, as related to the average, is 71 percent.

We tell you of these antics to prepare you, the owner, for the moment when you have to look at bid results, neatly tabulated by the architect's secretary. We don't want you to embrace the low bidder automatically, but to reflect on why construction bids often have these wide spreads. Besides the five classic components—labor, material, expenses, overhead and profit—there are additional factors, circumstances if you will, which influence a contractor when he composes the bottom line on the proposal to you. For instance:

► How eager is he to get the job—at this time, in this area of town, with a client like you, with the architect you hired? When he is too busy, his bids will be higher; if two of his next projects just fell through and he needs to keep his crew working, his bid will be lower. This has nothing to do with the cost of labor, material or overhead, does it?

► Another contractor is just starting out on his own. We hope he has enough experience and some working capital. His wife is his secretary, his truck is his office. He will lay out the building, build the forms for the footings, do most of the framing and can do the finish carpentry. A godsend. Just starting out and he will show the world what he can do. He wouldn't admit it but it is true—he has hardly any profit in this $22,560 remodeling job, just good wages. That is how much he wants his first project. His bid is very low. Send out a six-pack!

► There was another bid on the too-low side. This man is starting out also, except he does not know how to estimate; he forgot to include page two of the lumber list and guessed the cost of plumbing and electrical work. We found this out when we asked him for a construction cost breakdown. Yes, he has a license, but it is not really his, it is his father's. He would be a disaster, with or without a license.

There are more reasons for extremely high or low bids; we think we have given you a sampling. Be careful when you select a contractor. Expert advice is a must.

DO:
● Solicit enough bids to get a good feeling for the realistic cost range.
● Talk with the contractors whose bids seem extremely high or low. There is a chance that the high bidders have seen problems with the project all the others have overlooked. The low bidder might have missed something. Especially compare the allowances of the extreme bidders against allowances in the medium bids.
● If one bid is extremely low, look closer. You and your architect should check the construction cost breakdown. Compare the subcontractors' bids with those of the other general contractors. Don't laugh—does his math add up correctly?
● Catch "the new contractor on his way up."

So you think it costs more than you
thought. Welcome to the club.
Next let's talk about the costs.
There are ways to reduce them.

Help! The best bid is still too high . . .

B-10 A FEW SUGGESTIONS ON HOW TO LOWER THE COST

The lowest of four bidders is a small reputable contractor with good references. He works on only two or three projects at a time, which means that he will be on your jobsite almost full-time until the finishing trades take over. Further, he and his carpenter will not only do the foundation forming, framing and finish carpentry, but will install doors and windows as well and do all the pickup work. You rejoice. The ideal contractor, so far.

What's wrong? One thing. Even while his bid of, let's say, $80,000 is the lowest, it is still more than you were prepared to spend. Although the architect's estimated cost range was already higher than your dream price, you had hoped bids would come in around $70,000, maybe $75,000. And you know there are certain building-related costs that you have already paid and hoped to channel into long-term financing by making these expenses (survey, architect's fees, permits) part of the building loan. After the structure is completed, there will be bills not included in the general contract: carpet, window treatment, landscaping. What can you do to bring down the construction cost, to get the best bid lower yet, prudently, without changing the size or appearance or quality level? Consider a number of ways:

Be bold. If the size and quality level of your proposed building are exactly what you need and want, and what you would build, were it not for the qualms of the lender or your recently acquired habit of waking in the wee hours of the morning, if design and specifications are free of silly extravaganzas, if you hold a good job with a regular income and if, after a family powwow, all agree to forego the new car or the horse for a year and build instead, be bold, daring, walk on water—build it. The cost overrun should not be more than 10 or 15 percent, three or four years from now inflation will have compensated you IF you can hang in there.

How to calm the lender? There are intelligent devices for borrowing more than the bank thinks you should borrow. Use them, but only if you can live with high debts for a year or two:

In addition to borrowing from the bank, borrow from parents or against your life insurance; negotiate with the contractor to leave his profits in the project for some time and accept a promissory note; arrange for an interest-payment-only second trust deed or a second trust deed on which you don't even pay the interest (the interest becomes part of the balloon payment after three or four years). Have lunch with an experienced realtor or mortgage broker to arrange such a trust deed or mortgage. We have found one problem with some clients is not to dampen their appetite for building too much, but to caution them to not build too little. Avoid refinancing, however.

If the budget **must** be reduced, here are a few suggestions:

Build in stages. You can defer a portion of the project and build it later. The third bath, the fourth bedroom, a deck, a jacuzzi—all can be added later conveniently. Provide connections for the delayed building portion: footing ties, heating ducts, electrical junction boxes, plumbing capped at junction points. A better way to build in stages is to construct the entire shell, finish all exteriors and defer some finishing of the interior, provided the Building Inspection Department and the lender go along with the idea.

Make provisions for future installation of certain items or equipment: dishwasher, compactor, bar sink, intercom, built-in vacuum cleaner, bookshelves, window screens—just provide the space and utility connections where needed. We talked about that in Chapter A-14.

Consider less expensive alternatives. Carpet in lieu of wood flooring, a fiberglass shower in lieu of ceramic tiles, white plumbing fixtures in lieu of colored ones, etc.

Conduct a second round of bidding. Inform the contractors who have submitted bids that you cannot accept any of the bids. Start all over and invite a new slate of general contractors to bid. Seldom do we recommend this procedure. Experience has shown the contractors of the first round probably will not bid again nor lower their bids. To assemble a new list of good contractors, to plow through the bidding spiel again, to evaluate the new bids, will be time-consuming. During the delay inflation could eat up any gain you may make.

Be your own general contractor, sometimes called owner/builder. Theoretically you save all the money the general contractor has in the deal for expenses, overhead, risk and profit. *Theoretically!* Seldom has an owner the time and the patience, let alone the experience to get bids from subcontractors and coordinate successfully even the smallest construction project. As an educated guess, thinking back to the botched-up, loused-up, costly, snail-pace owner/builder projects we have seen, in about one out of 20 cases the owner comes out ahead. One out of 20, and we are talking about small projects. We don't recommend being your own contractor. There is no substitute for an experienced contractor. At least hire one to build the shell, then you finish the building yourself, or have it finished by subcontractors you hire. But even this we don't recommend often—have the owner do the entire finishing on his own, that is.

This brings us to three additional methods to reduce costs; all are proven and **could** bring the cost below the best bid submitted. Because these methods are complicated, we'll deal with them in separate chapters:

After reading this chapter and the next three you will find, we hope, a solution to your particular problem. The solution may be a combination of the methods explained. You might:

Be bold to a certain limit.
Defer a portion of the building.
Make provisions for equipment to be installed later.
Decide to go "cost plus" in lieu of a lump sum contract.
Finish a part of the project yourself as owner-builder.
Interchange some subcontractors or invite additional bids on a few subcontracts.

The message:
There are devices to get the best bid still lower without changing the size, appearance, quality level or timing of your project. Study and understand these methods.

DO:
- Sit down with your architect and the selected contractor and discuss the options.
- To get the best bid lower, consider selecting a **combination** *of cost-saving methods.*

Another method to lower the cost . . .

B-11 CONSIDER A "COST PLUS FEE" CONTRACT

Consistent with sound business practice, to arrive at the lowest possible construction cost, we have emphasized the need for
precise construction documents
a selective list of contractors interested in bidding
competitive bidding
a high degree of the owner's involvement during all phases.
Then the time comes to sign the construction contract form, the Agreement. The type used in most cases is the so-called "lump sum" contract, also called "stipulated sum" or "fixed price" contract, resulting either from negotiation with a single general contractor or from competitive bidding among several general contractors.

The lump sum contract has the advantage of simplicity: it has a bottom number; the owner knows what he has to pay for the project. It also has a certain flexibility: changes in the work can be made either by using specified alternates or by writing change orders. It also has distinct disadvantages:

- The bottom price, rather than the competence of the contractor, becomes the overriding reason for contractor selection, even if all general contractors are looked over rather carefully before being invited to bid. The owner's constant task is to make the contractor perform per contract.

- What happens when the market place fluctuates after the contract is written? Will sudden shortages delay the project or make the owner dip into his savings account to buy gypsum board from Timbuktu or Polish nails from New Zealand? Who takes the windfall when lumber prices drop during construction?

- It is difficult to add to or deduct from the contract, get proper credit for deletions and pay a fair price for additional work since such work is really not bid competitively anymore; the subcontractors you have on board at the time of the bid date will do the work.

As clearcut as the typical lump sum contract seems to the layperson, most construction experts agree there is a better type of owner-contractor agreement. One which retains the advantages of competitive bidding and fairly projects construction costs, but also increases flexibility to add, change or deduct, even to the point of making design or material decisions during construction. One which also makes the owner pay all contractors the true costs as close as possible, reverts cost savings and windfalls to the owner and eliminates the often justified general contractors' fears of not making a profit. This type of contract guarantees him a fair profit, a matter in which the owner should be interested. It is called the "cost of the work plus fee" contract and is a most economical, satisfactory procedure, IF executed by a reputable general contractor.

The steps leading to a cost plus contract are similar to those that will produce the typical lump sum agreement, especially the first three:

► Get precise drawings and specifications, either good preliminaries or final contract documents.

► Invite three or four reputable contractors to arrive at a tentative construction cost range, sometimes also called "probable construction cost."

► Ask each general contractor to quote a fixed fee to build the project. The fixed fee takes care of his home office overhead, risk and profit.

► Select a general contractor on the basis of:
   a. competence
   b. fixed fee amount
   c. tentative construction cost range with an upset price (also called a guaranteed maximum price)
   d. construction time

► The construction cost breakdown of this general contractor then might consist of four categories:

—Work performed by general contractor's own employees (probably: layout, foundations, rough carpentry, finish carpentry)

—Subcontracts

—General contractor's jobsite overhead (field office, superintendent, utilities, etc.)

—General contractor's fee

Work performed by the general contractor's own forces and jobsite overhead must be substantiated by itemized pricing, time sheets and material bills.

Work performed by subcontractors will be bid competitively; a minimum of two proposals should be submitted to the owner from each trade, vendor, equipment supplies, etc.

Only the fee is fixed and can be increased only if the scope of the total work increases by more than 5 percent.

► You can fine-tune the arrangement: Rather than selecting the general contractor when the probable construction cost is established, wait with the selection and let three or four general contractors submit firm bids in a competitive manner. Request in the instructions to the bidders that each contractor not only proposes a lumpsum but also state what his fee would be if the owner decided to choose a cost plus fee agreement in lieu of the lump sum contract. Then, on the basis of competence, fixed fees and lumpsum bids select a general contractor. That contractor's lumpsum bid will then be accepted as the maximum contract cost of the cost-plus-fee contract.

Even while under construction, the general contractor keeps accepting subcontracts in a competitive fashion. Any savings which will bring the cost under the agreed maximum will be divided between owner and contractor according to a formula, say 35 percent of the savings remain with the general contractor, 65 percent will be passed on to the owner. This gives the contractor an incentive to massage down the costs.

DO:

- Consider a cost plus fee contract. You need professional help to draw up the agreement. Award such a contract only to a general contractor who fully understands the procedure and has operated continuously with cost plus contracts.
- Use A.I.A. Form A-111 (the title page is shown on the next page). Have in writing what the "costs" are and what the fee covers.
- Be prepared for extensive time-consuming record keeping, and for review of subcontractors' proposals during construction.
- Reserve the right, as owner, to have separate contracts and also to have certain building-related costs paid from loan proceeds.
- Observe all prudent rules pertaining to any construction contract: Have general and special conditions, guarantees and contract administration by an experienced professional. Keep track of unused portions of allowances. Conduct a one-year-after inspection.

DO NOT:

- Confuse a cost of the work plus fee contract with "time and material." **T & M** means you tell the general contractor, in effect, build it and send me a bill! No competitive bidding, no construction cost projection, no guaranteed maximum cost. With **cost plus fee** the general contractor's office overhead, risk and profit are predetermined, as is the cost range of the total construction. A guaranteed maximum can cap the range.

Once more: The cost of the work plus fee contract involves a lot of administrative work during bidding, contract negotiation and construction. It is worth it.

The frontispiece of the Cost Of The Work Plus A Fee Agreement is shown on the next page.

THE AMERICAN INSTITUTE OF ARCHITECTS

AIA Document A111

Standard Form of Agreement Between Owner and Contractor

where the basis of payment is the

COST OF THE WORK PLUS A FEE

1978 EDITION

*THIS DOCUMENT HAS IMPORTANT LEGAL CONSEQUENCES; CONSULTATION WITH
AN ATTORNEY IS ENCOURAGED WITH RESPECT TO ITS COMPLETION OR MODIFICATION*

Use only with the 1976 Edition of AIA Document A201, General Conditions of the Contract for Construction.

This document has been approved and endorsed by The Associated General Contactors of America

---

AGREEMENT

made as of the   day of   in the year of Nineteen
Hundred and

BETWEEN the Owner:

and the Contractor:

the Project:

the Architect:

The Owner and the Contractor agree as set forth below.

AIA DOCUMENT A111 • COST-PLUS OWNER-CONTRACTOR AGREEMENT • NINTH EDITION • APRIL 1978 • AIA®
© 1978 • THE AMERICAN INSTITUTE OF ARCHITECTS, 1735 NEW YORK AVE., N.W., WASHINGTON, D.C. 20006   A111-1978   1

To get the best bid still lower . . .

B-12 FINISH THE BUILDING YOURSELF?

Generally speaking, a building is constructed in two operations; one follows the other, with some overlapping:

Erecting the shell: Grading, foundations, utility lines, slabs and floors, walls, roof, windows, exterior doors, rough plumbing, electrical and heating, exterior plaster or siding, exterior trim.

Finishing the shell: Insulation, drywall or interior plaster, paneling, cabinets and pullmans, interior doors, tiles, finish carpentry, flooring, painting, final grading, walks and patios, clean-up.

To lower construction costs: Hire a general contractor to erect the shell of your building. Then you, as owner/builder, let contracts directly to subcontractors, who finish the shell. This method works well in cases where:
a. The project is small (residential remodelings, houses, commercial additions of limited size).
b. The owner has time, patience and an inquisitive mind, is a practical person and has some experience with construction people.

   or

c. The owner is very experienced and intended to do the entire project as owner/builder, but was advised by the lending institution that as a condition for granting the loan, a general contractor has to be on board and be responsible for the construction. The owner then has to play it by ear—how much will the lender let him do as owner/builder and to what degree must the general contractor be involved? Most lenders have had their share of botched-up owner/builder projects. If the lender allows it, or better said, does not object, identify yourself on the loan application and the cost breakdown sheet as owner/builder and the general contractor as subcontractor building the shell. If this arrangement gives the lender the willies, identify the general contractor as responsible for the entire building and then make arrangements with him as to which trades you will handle with full responsibility. Keep the lender informed; he probably will want a copy of your agreement with the general contractor anyway. When responsibilities are split up, it must also be decided who writes the vouchers for fund control.

If contracting the finish work appeals to you as a cost-saving device, we still want to caution: You will not save all the general contractor's overhead and profit on those finishing trades (for reasons explained in Chapter B-10, where we advised against trying to be your own general contractor). In most cases there is no substitute for an experienced contractor!

How much work the general contractor should do and how many trades you as owner/builder should handle depends on how much time you have, how able you are to adjust to the antics of the construction business and how willing you are to learn fast and be patient, thorough, hard-nosed and a few other things all at the same time. Be that as it may, which trades could an owner/builder handle after the shell is constructed?

Probably: painting, carpet installation, cleaning, sprinklers, landscaping
Maybe: ceramic tile work, resilient flooring, cabinets, light fixtures, pews

We would not recommend much more, even under the best of circumstances. Just so there is no misunderstanding—we aren't implying you should do all this finishing work with your own two little hands. You let the contracts for the trades mentioned, provided you can read drawings, understand the specifications and know how to negotiate. You become an owner/builder for a portion of your project.

When you become an owner/builder you will not only take on management duties, you will take on some legal responsibilities. If:
• "you employ or otherwise engage" any person other than your immediate family.
• the work (labor, materials, expenses) is $200 or more for the entire project, and
• such persons are not licensed as contractors

then you may be an employer. Legally you become the subject to several obligations such as state and federal income tax withholding, social security taxes, workers' compensation insurance, disability insurance and unemployment compensation contributions.

Hiring a licensed contractor to build at least the shell takes care of a part of the burden; this arrangement might save you grief and costs. But no matter how much chance an owner wants to take in regards to the tax ado, he would be more than foolish to skimp on insurance for the workers he employs directly.

Summing up:
In case the cost-saving method of the
▶ shell-by-a-contractor
▶ finishing by-the-owner
appeals to you, here are the DOs and DON'Ts:

DO:
● Try to find a contractor who specializes in shell construction. You might get names from the local Building Contractors Association or from advertisements. You could place an ad yourself under "Contracting," "Contractors & Builders," or "Help Wanted" (See B-2.)
● Have precise contract documents with the shell contractor, as you would for a complete project, namely: an agreement, drawings, specifications, general and special conditions and, during construction have written change orders, professional inspections, punchlists, guarantees.
● Be sure drawings and specifications delineate clearly what is part of the shell contract and what is "finishing" as handled by the owner. Again, the agreement should

126

enumerate trade by trade the work to be done by the shell contractor.

- Anticipate! Visualize the finished surfaces. For instance, where the finish will be mortar-set ceramic tiles, the shell contractor must provide a scratch coat, not gypsum board. The contractor must know this at the time he bids the shell.
- Realize that certain trades will work in both phases, shell and finishing: Rough plumbing is part of the shell; setting toilets, lavatories and faucets is finishing. Electrical panels and conduits are shell; light fixtures, switches and wall plates are finishing. Heating ducts and furnace are shell; grilles are finishing. Leave these trades entirely under the shell contract, with the "finishing" portion of each trade to be installed by the same subcontractor who worked under the "shell" phase.
- When handling your end of the project, ask the general contractor for advice and names. Most likely he will gladly help you.
- Obtain a checksheet on the legal aspects of being an owner/builder. Your local building department should have one; if not try the nearest U.S. Small Business Administration office or even the I.R.S.

DO NOT:
- Believe farming out the finishing trades is easy and automatically yields a bundle. Depending on your efforts it **can** save money. And you must have time, time, time.

To get the low bid lower yet . . .

B-13 INTERCHANGE SUBCONTRACTORS

You have gone through the pandemonium of competitive bidding and from a selected list of good contractors have received three bids:

A $44,850
B 39,700
C 36,250

You have accepted C as the best bid and are ready to sign a contract. As a prudent person, you ask yourself and your spouse, of course, asks you: Besides competitive bidding, are there other methods to get the low bid even lower? We have mentioned already:

- **Build in stages.** Make provisions, use less expensive alternates, go through a second round of bidding, be an owner/builder. See B-10.
- **Cost plus fee.** See B-11.
- **Finish the building yourself.** See B-12.

Anything else? Yes. Interchange subcontractors. It is a time-consuming method, obviously cumbersome and therefore not often used. It is also not popular with general contractors.

How does it work?

In your instructions to bidders you requested from all general contractors a list of the subcontractors they intend to use if awarded the contract. That was not a favor you requested, but a condition of the bidding. Some general contractors might state that a list of subs becomes available to you after he is assured of getting the contract. This type is the exception; don't fight it. Most generals will list their subs if you request it.

Assessing the best bidder's proposal leads you to assume that he also succeeded in assembling the lowest-bidding subcontractors in town. This is not necessarily so. By contacting the subcontractors of the other two general contractors, you find that some subs are lower than the subs your general contractor listed. Take the three electricians. You phone and find out that one electrician is not interested in giving his sub bid since his general contractor did not get the contract, but the other electrician says his bid was $2,140. On your general contractor's cost breakdown sheet, you find $2,500—a difference of $360. If you and your contractor can pick up two or three or four lower subcontractors from the other list of subcontractors, you may save $1,000 or more. That's another 2.76 percent:

$$\frac{(\$1000 \times 100)}{\$36,250}$$

Your general contractor shouldn't mind doing this "bidding after the bidding" since it benefits not only you but also him—he should be glad to find another low subcontractor whom he can use on future jobs.

Interchange subcontractors—why isn't it done more often?

► Subdivision or apartment developers, who are both owners **and** general contractors, interchange subcontractors on every project. Their hard costs (amount of all subcontracts, equipment and other field costs) are the result of intelligent hard-nosed shopping among subcontractors and vendors.

The average private owner, however, who has shopped among three or four **general** contractors, is understandably under the illusion that the most favorable bid he is willing to accept from one of the general contractors is already the result of vigorous competitive bidding among each trade group of subcontractors. The owner assumes the "best bid" is just the summation of all the lowest subcontractors' bids plus the general's overhead, risk and profit.

In most cases, this is not so. General contractors have their favorite subs—people they have worked with in the past, whom they know and trust and who have delivered low bids before.

► Interchanging subcontractors is not popular for another reason: General contractors also know subs they do **not** want to work with, for real or imagined reasons. It follows that a general contractor does not see much purpose in shopping around after he has been selected as the best bidder—he would be trading a known sub for an unknown and might even expect to lower his profit margin, which was based on the sum of the subs he had submitted. So why spend the time to invite problems?

As owners we would (1) agree to do some of the afterbidding ourselves, if the general contractor declines, and (2) leave the original profit margin for the general contractor unchanged, adjusting only the sub-bids to arrive at a lower total cost.

► The other two general contractors might object to contacting their subs, although there is no legal justification. If they strongly object, don't contact the sub. You and your general contractor can still find enough subcontractors if you want to keep bidding. Look through the published bid results in the **Daily Construction Reporter;** often the names of subcontractors of other successful general contractors are listed.

► Interchanging subcontractors involves additional work including many phone calls and sending out prints. Don't print new sets of drawings and specifications, however; use the ones returned at the bid opening. The additional work can be done by the general contractor or an alert owner. Let's say it takes you 10 hours, for example, and you shave $1000 off the $36,250 contract. That's another 2.75 percent. $1000 divided by 10 hours . . . my gawd, that's $100 per hour—more than you make down at the Plating Plant!

► Subcontractors or suppliers are interchanged but the owner is not aware of it. A few general contractors do it after the construction agreement with the owner is signed. It is called "bid shopping." But after the shopping the contract price will not be changed. Pssssst—the cost savings such as the electrician's $360 will be pocketed by the general contractor. Fortunately, it does not happen often.

What can the owner do? Insist on a list of subcontractors with the bid and insist on an explanation when names of subcontractors are changed.

When we, for the benefit of the owner, interchange subcontractors, we use an incentive for the general contractor—the same method we use with cost plus fee contracts: we not only leave the general contractor's profit margin unchanged but also split with him the cost savings—35 percent to him and 65 percent to the owner.

DO:

● Discuss with your general contractor interchanging subcontractors. A prerequisite: all bids received must include a list of subcontractors.

In summary, let's review the last four chapters. In B-10 we realized that the best bid was still too high and we looked at ways to bring the costs down. The solution in your particular case might be a combination of the methods explained:

Be bold. Increase your budget somewhat.

Build in stages; defer a portion of the project.

Make provisions for some equipment to be installed at a later date.

Finish a portion of the project yourself or under separate contracts.

Consider a cost-plus-fee contract.

Interchange some subcontractors or solicit additional bids on some subcontracts.

Discuss with your architect or general contractor the options available—then use any of these methods. Then determine the contract price. 'Course it will be lower than it was three chapters ago!

The Owner-Contractor Agreement must
reflect all your efforts of the previous
months to build better—faster—for less.
Study it carefully. Signing the agreement,
the "contract," is next.

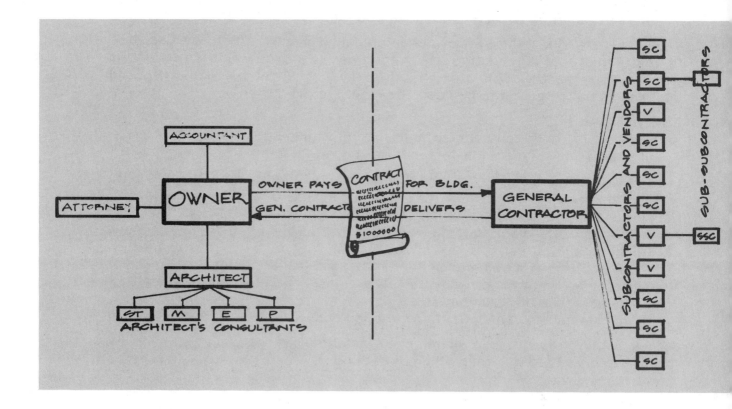

The typical OWNER - GENERAL CONTRACTOR relationship

Main points:
- **Owner** and General **Contractor** are the two parties to the typical building construction contract.
- The contract has several parts:
  - The Agreement
  - Drawings
  - Specifications
  - Addenda
  - Change Orders

  and any document specifically designated as a part of the contract.
- Subcontractors, material suppliers and vendors have a contractural relationship with the general contractor but not with the owner. Except when they have a separate contract with the owner.
- The **architect** and his consultants are not parties to the typical building construction contract, but the architect has a separate contract with the owner. *Exception:* Sometimes the architect is the owner. But then his consultants, of course, don't have a contractural relationship with the general contractors.
- When the owner decides to be an **OWNER/BUILDER**, then all subcontractors and vendors contract with him directly.
- If interested in any more meaningful relationships, please, watch *The Paper Chase* one more time.

132

B-14 SIGNING THE CONSTRUCTION CONTRACT

After selecting the general contractor and before signing the construction contract, the owner, with his architect, might want to review drawings and specifications one more time. If either one wants changes they should be discussed with the general contractor now, before signing the contract. Changes **during** construction will most likely incur additional cost; now a change can be handled by an addendum and an adjustment of the contract sum. Often contractors are generous at this stage and will "throw in" this or that, a generosity which will dissipate as soon as construction is under way.

In most states, in order to be valid, a construction contract must be in writing, dated and signed by both parties. It is best for the owner to use the latest standard contract forms of the American Institute of Architects, which have proven to be practical and also equitable to both parties; most A.I.A. contract forms are endorsed by architects and contractors associations.

There are several forms, all called "Agreement Between Owner and Contractor," with each designating a specific type of contract: "Stipulated Sum" (lump sum) or "Cost of the Work Plus Fee" or "Cost Plus Fee With a Stipulated Sum" (a top set sum). The usual contractor proposal forms you buy at an office supply store are not accurate enough and are slanted to the advantage of contractors.

Not only the agreement, but **all contract documents,** although some don't have to be signed, **become your contract with the contractor.** It legally consists of these documents: the agreement; drawings and specifications, numbered and dated; and any addenda issued during bidding, numbered and dated. If during construction change orders and X-drawings will be issued, these also become part of your contract. After the contract is signed, the owner has three working days to cancel it.

Relay the outcome of the bidding and your selection of the contractor to all general contractors who had submitted a bid. It is a matter of courtesy; you do this by phone or by sending a short note.

"We thank you for your efforts to submit a proposal for our new residence. With our architect we have selected So and So Construction Co. as the best bidder. We certainly will recommend you to our friends . . . etc."

If one of the contractors you notified was not at the bid opening you could, as a courtesy, attach the tabulated bid results.

If there is the slightest chance of a deferred construction start or if you plan modification of drawings or specifications, be careful not to send to the selected contractor a Notice of Award or Letter of Intent. Although rare, a legally enforceable contract could arise between owner and contractor from such notice or letter. Consult your attorney if you have doubts about formal letters to the contractor or about the recommended contract forms, especially if you or your architect intend to change the forms. The contract forms interact with the "General Conditions, A.I.A. Document 201," which definitely should be a part of your contract as a division of your specifications. If you have any partnerships, a trust or other legal arrangements, your attorney might advise you as to the effect of your construction contract on those arrangements.

DO NOT SIGN THE CONSTRUCTION CONTRACT BEFORE:
- You have a building permit.
- Your construction financing is assured.
- You have checked out the contractor.
- An architect or another professional construction expert has reviewed all contract documents and has found them in order and of latest issue, numbered and dated.
- You have decided whether to bond or not to bond the contractor.

After the contract is signed and before construction starts you want a "Certificate of Insurance in Force" from your general contractor and from any other contractor not working under the general contractor's insurance. If you intend to hire workmen directly, obtain Workers' Compensation Board Insurance and Liability Insurance; contact your insurance adviser on this matter. As soon as the framing work starts, increase your fire insurance; whatever is nailed, bolted or even braced to your land is, from now on, your responsibility if fire breaks out.

Give the permit set—the one with all those stamps—to the general contractor; he should keep it on the jobsite, inside the building, under a protective cover and hidden. The Building Inspector should be informed of the location of the permit set in case he comes for an inspection and nobody is on the building site; a rare occasion, but it happens. It should not be given to any workman or to any vendor or salesman to make take-offs. It is more than a pain in the neck to have a permit set replaced once someone has taken off with it.

Your contractor also needs additional sets of prints, normally six to 10, for his workmen and subcontractors. Be sure he gets the latest prints. Do not allow any outdated prints to float around on the jobsite. No outdated prints! Ever! Burn them the night you sign the contract. Light the fireplace with them before you open the special bottle of Pinot Noir to celebrate the event. A great moment. You and your spouse, the family, the Building Committee or business partners—whoever decided to get something built—are getting closer. Monday morning construction will start.

No, you didn't forget! Someone had reminded you to have the contractor bonded. The following chapter tells you whether or not you need a bond.

B-15 GET A BOND! WHAT BOND?

Relax. You probably don't need a bond. Of the last 40 or 50 projects our firm was involved with, none had a bond, from the proverbial add-a-bedroom-and-bath to the $280,000 residence to the $2.8 million shopping center. Still, in case your architect or attorney brings up the subject of bonds, you might want to have some basic information.

What types of bonds are used in construction?

The full description is "surety bond." It guarantees that one party (the "surety") pays or makes good to the other (the owner or client, also called the "obligee") for losses that result from failure of a contractor (the "principal") to perform per contract. There are several types of bonds, all of which are issued by a surety company, also called a bonding company. The fee is paid by the owner, the obligee.

Bid bond: Surety pays the owner if a bidder fails to enter into a contract with the owner within the time specified in the bidding instructions or on the proposal form, normally 30 days. The surety will pay either
a. the amount of the bid bond   **or**
b. the difference between the "best" bid the owner had selected (and then the bidder failed to enter into a contract) and the higher amount of the next available bidder.

Performance bond: If the contractor defaults, the surety will pay the owner losses up to the bond amount and/or arrange and pay for completion of the project, including any additional work necessary by architects, consultants, inspecting agencies, another contractor, etc.

Labor and materials bond: Surety will pay off a contractor's obligations for labor, material, equipment and tool rentals, if the contractor was already paid by the owner for these items but did not pay **his** bills.

There are other bonds, but they are rarely used. Check with your architect on these: completion bond, lease bond, license bond, roofing bond, termite bond, subcontract bond.

HOW DO YOU ARRANGE FOR BONDS? WHAT IS THE COST?

The contractor should contact a bonding company or an insurance agent, furnish all required information and note the amount of bonds and costs on his proposal form. The contractor can find the names of bonding companies from contractors' associations, attorneys, architects, independent insurance agents specializing in commercial property and liability insurance. Or in the yellow pages under "Bonds—Surety and Fidelity."

Cost of bonds: About .75 to 1½ percent, sometimes up to 3 percent of the bonding sum, depending on the rating of the contractor by the bonding company. Since the owner receives the protection, he pays for the bonds, either separately or as part of the building costs. The cost of bonds should be noted on the contractor's proposal form and also on the cost breakdown sheet. Bond cost can be channeled into long-term financing unless the owner's accountant or attorney advises against it. The owner has the right to approve the bonding company.

Forms: Most government agencies have specific bond forms; private owners are advised to use the A.I.A. bond document. Whoever initiates the purchase of bonds should take advantage of the services of a competent agent who specializes in construction insurance bonds.

Bonds are needed when:

- The owner is a government agency or an association obligated by law or by-laws to acquire bonds.
- The bid of the selected contractor seems unusually low, but owner and architect decide to accept the bid; or the contractor seems competent but overextended.
- Only one bid was received and no comparison is possible.
- Price rises or labor troubles loom and yet the owner does not want to enter into a cost-plus-fee contract.
- There is a reason for qualms, but not a strong enough case to drop the contractor: sickness or death of a contractor or key personnel; contractor engages in speculative ventures elsewhere; apparent lack of working capital; tax problems; etc.
- You, your architect and your attorney won't sleep well without having a bond.

Summing up:

Bonds add to the already high cost of construction. And it must be said again: No bond, no contract is a substitute for the careful selection of an experienced contractor with integrity, financial staying power, adequate work force and an agreeable personality.

We have also found that some contractors have difficulty getting bonded, for different reasons. Some despise the paperwork and do not furnish enough information to the bonding company; some plainly can never be bonded because they are—often successfully—into so many projects that their lack of working capital gives the bonding companies the willies. Other contractors bid so low that by the law of economics they cannot stay in business for long; at least the bonding people would think so.

Bonding can take from one to six weeks, depending on whether or not the contractor had qualified for bonding before. The initial work in establishing a bonding service is very comprehensive. In checking out financial statements, track record, banking relationship, and other underwriting points taken into consideration, the surety uses up a fair amount of time.

Some contractors notoriously postpone submitting information to the bonding company until the last days before the bidding deadline. This prevents the underwriter from checking out the information and committing the bond on time. And some contractors do not qualify by the surety company standards of the famous "three Cs"—credit, capital and character—about which some of our best contractors laugh all the way to the beerhall. On one hand, then, the prequalifying requirements of the bonding companies might leave you with a meager but top-of-the-crop list of good contractors; on the other hand the same standards eliminate from your list some contractors who have the experience, guts, dedication to workmanship, time schedules and stretched financing to build your castle! What, then to do?

Since most projects legally do not need bonds, you probably have a choice. Consider bonding because it gives additional protection. It is not a substitute for selecting a good contractor, however. There is a smart compromise. In instructions to bidders, insert a note that all bidding contractors "should be bondable," in case bonds are desired by the owner. **Bondable** is the key word. It implies to the contractors that certain standards are required, even if final proof of bonding capacity is not expected. If a contractor cannot or does not want to be bonded, he should be asked to explain the reasons, yet still submit the bid. The "should be considered bondable" clause screens without eliminating the occasional exceptional contractor who has more experience, capability and character than capital.

If a contractor cannot be considered bondable but delivers a good proposal, has references and is eager to do the project, consider giving him the contract but definitely take all these precautionary steps:

- Verify the references, one by one
- Have precise contract documents (drawings, specifications, addenda and a professional agreement)
- If the contractor is incorporated, sign the agreement with him as a corporation and as an individual
- Don't pay any advances
- Have fund control
- Obtain written guarantees, from the general contractor and from the subcontractors individually
- Hire professional help to oversee the construction and conduct final inspections
- Get material and labor releases
- File Notice of Completion
- Follow certain steps before making final payment (See chapter C-18)
- And hope

Authors note: This form is carried by most good stationary stores throughout the West.
W.R.H.

RELEASE

MATERIALS AND LABOR

Date...

TO WHOM IT MAY CONCERN:

For a valuable consideration the undersigned..hereby releases the property

at.., in the City of ....,

California, from any liability for lien for all materials delivered by it, and labor performed. to or for that said property for or on account of

.... to this date.

This release is conditioned upon the clearance by the bank upon which it is drawn, of the check received in payment for the above mentioned materials and labor.

_____

_____   _____
                                                                          FIRM NAME

_____   _____
                                                                          ADDRESS

_____   _____
                                                                          CITY

Courtesy of
Wolcotts Form 1514—Rev. 2-60   _____
                                                                          AUTHORIZED REPRESENTATIVE

Typical Materials and Labor Release form (Waiver of Liens)

It will help you now to read the Cost-Shaving Checklist on the following pages. Check your project against every recommendation on the list.

3-16 COST-SHAVING CHECKLIST
PART B (BIDS, FINANCING AND THE CONSTRUCTION CONTRACT)

Refer to	DO	Importance to reduce construction costs, hazards or liabilities	Pertains to your project	Done
B-2	Assemble a list of good contractors.	■ ■		
	Try to catch "the new contractor on his way up."	■ ■		
B-4	Bid and build when the other guy isn't.	■ ■ ■		
B-5	Obtain at least three bids — on almost everything.	■ ■ ■ ■		
B-1	Shop resolutely for best financing.	■ ■ ■		
B-8	Observe certain bidding procedures. State bid date, time, place, list of bid documents, bonds required, deposit, return of documents.	■		
A-9, B-5	Control rigorously building-related costs: Survey, reports, professional fees, financing, permits, Fund Control, escalation clauses, insurance, utility costs, testing, inspection fees. Channel many of these costs into the building loan.	■ ■		
B-6	Encourage contractors to suggest cost-saving alternatives.	■ ■		
B-8	During the bidding period: use inquiries from bidders to eliminate most bugs from drawings and specifications. Issue ADDENDA to answer contractors' questions.	■		
B-7	Specify that unused portions of ALLOWANCES to be returned to owner prior to final payment.	■		
B-10	Bids too high? Consider: to build in stages	■		
	a second round of bidding	■ ■		
	your own list of cost-reducing alternatives	■		
	provisions for future installations	■ ■		
B-11	"Cost-Plus-Fee" in lieu of "Lump Sum" Contract	■ ■		
B-12	Separate contracts	■		
	finishing the building as "Owner/Builder"	■		
B-13	interchanging subcontractors	■ ■		
B-14	Use A.I.A. standard contract forms	■ ■		
B-1	Have financing assured before signing construction contract	■		
B-14	Make sure that the construction contract form (AGREEMENT) enumerates: number and date of drawings, specifications, addenda and/or any other document which becomes part of the contract	■ ■		
Refer to	**DON'T**			
B-1	Agree to pay Dutch Interest.	■ ■ ■		
B-9	Allow any contractor to submit a bid without a detailed, trade-by-trade, cost breakdown	■		
A-17, B-14	Consider bids final until all necessary corrections to drawings, specifications and addenda are made and acknowledged by bidding contractors.	■		

PART C: CONSTRUCTION PHASE

If there is a time for every purpose under heaven,
and to everything there is a season,
then there is—probably—a . . .

C-1 BEST TIME TO BUILD

When, in general, is the best time to build?

When you have sufficient time for planning, arranging financing, bidding and reviewing drawings and specifications, in most cases with professional help. It is easier to change a wall on the drawing than in the field; it is easier to save $100 or $1000 through skillful negotiation with a lender in the bidding and financing stage than to earn the money later to pay off an outrageous loan for years to come. Allocate sufficient time.

When interest rates are not at peak. On the other hand, since interest rates and points always seem higher than we think they should be, don't wait and wait, hoping for interest rates to come down dramatically. During the last three years, construction costs climbed about 13 percent per year; completed buildings appreciated at even higher rates. Therefore, you have to weigh the cost of the construction or building loan at any given time against the cost increases by inflation and appreciation (appreciation you didn't get because you didn't build). The high cost of money brings us to another suggestion:

Build at a time of low building activity, which often coincides with the time of lowered interest rates. To put it boldly: build during a recession! That takes guts and a sense of timing, because you must have your drawings, specifications and permits ready when the economy is on the downslide. Perhaps your time schedule does not allow you to wait that long. But we think it is sometimes more a matter of guts than the necessity to have the building in a particular month or year, especially with residential or church remodelings.

Building during a recession, or an economic downturn, does not sound so alarming in view of the following:
- You get better interest rates and loan fees.

- More contractors are eager to bid (even plumbers return your phone calls); you will have more bids to choose from.
- No hassle with material shortages.
- Probably no labor disputes.
- Workmen produce better due to a more competitive job market; you can even make a few suggestions on the jobsite!

Therefore, in capital letters, to individuals, to church building committees and to corporate boards, our most valuable advice:

BUILD WHEN THE OTHER GUY ISN'T!!

After selecting a contractor, you sometimes can get an even better proposal if you time the construction, to a certain degree, to the general contractor's convenience. This does not mean: This year, if he has time, otherwise, next year! It means: Let the man finish the project he is working on right now, so he can move his men smoothly, maybe in six weeks instead of two, from that job to yours.

When to break ground, to actually begin construction? Don't begin until:

- Escrow has closed (in case you bought a lot or had to sell a building first).
- The building loan is approved in writing and recorded and the lender has sent you a written Notice to Proceed. The title company probably will take a picture of your building site in the same hour the loan is being recorded. Don't have a stick of lumber on the site unless you have agreed with the loan officer beforehand, and IN WRITING!
- You have a written contract with your contractor or, in case you are owner/builder, with your subcontractors.
- You have received your contractor's certificate of insurance, with an expiration date beyond your building completion.
- You have a building permit, if needed; arrangements for water, electricity and job toilet have been made.
- You have received clearance from the Architectural Control Board, if there is one.
- You feel reasonably sure that the work schedule allows you to have the roof on before the rainy season starts.
- Final suggestion: Be sure you are building on the correct lot—yours, not your neighbor's. In certain new subdivisions, one lot looks like the next. If you think this is funny, read some of the not-so-funny court proceedings on the subject. That someone would build on the wrong lot sounds ridiculous, but it has happened. And to straighten out that mistake is a nightmare.

Check every item above.
All Affirmative?
Call in the troops.

C-2 WHERE ARE THE TROOPS!

Once construction has started, keep up the momentum. Workmen should be on the jobsite full-time almost every day for a number of reasons:

Time limit of permits.

We mentioned permit expiration dates in Chapter A-19. But since so many snafus develop because some owners don't pay attention to those small notes on permits, let's review them once more at this critical stage.

From the date of issue of the building permit, you have only so many days to call for the first inspection, that of the footing excavations. Where the Uniform Building Code governs, it is 120 calendar days. Roughly four months is enough time to grade; dig trenches; install rough plumbing, electrical and heating; form the foundations; lay reinforcing steel, and have the site ready for inspection. But we have seen cases where all this work was not done within the allotted 120 days. Why? Bad weather, financing not finalized at beginning of construction, material shortages, labor disputes or the owner or contractor just diddled around. So, the building permit expires automatically, by statute, when the foundation or any other first inspection is not conducted within 120 days after issue. The Uniform Building Code provides no legal means for an extension. Maybe a friendly Building Inspector, by not looking too closely for the expiration date on the permit, will (illegally) allow a few days' grace period.

Once the building permit has expired, you have to renew it by paying an amount equal to **one-half the building permit fee** again, although not the plan check fee. Some years ago, one of our clients had a permit for a hotel building. His construction financing ran into a snag, the 120 days passed and he again paid half the permit fee of approximately $2,400 to revive the permit.

Sometimes the renewal of a permit becomes more complicated, as when the building permit is tied to a **Conditional Use Permit** or a **Variance,** which by their nature have time limits. You may end up having to apply for these special permits again, which could mean more public hearings and waiting time. Another time-consuming scenario can develop when the permit expires and the Uniform Building Code has changed during those 120 days. In order to renew the permit, the building must conform to the new code. The UBC normally is revised and newly issued every three years. The latest codes were issued in 1973, 1976, 1979 and 1982.

The rain

falls not only on the plain in Spain, but, with heavenly delight, on buildings under construction. Especially residential remodeling projects where the roof has just been opened up to the bare ceiling in order to receive the proposed second story! Anticipate the rainy season and schedule your building accordingly. On remodeling projects where part of the buidling is inhabited during construction and building activity extends into fall, have enough 6-mil Visqueen handy and 2 x 6's to weigh the edges of the sheet vinyl down should you try to cover the openings in the roof above your loved ones in the middle of the night.

There is another advantage to getting the roof installed as soon as possible: Exterior finishes, windows, doors and locks can be installed. The building should be locked up as soon as possible. Only after this can you obtain insurance coverage against theft. Vandalism and theft are enormous and annoying problems in construction. People steal what they can load on a truck from entire lumber stacks delivered only yesterday to Heatilator fireplaces to kitchen disposals. It is almost impossible to purchase insurance against theft while the building is open. And open and unlockable it will be until the roof is in place.

Construction loan interest

is paid by the owner for every day, every week, every month the building is under construction. Keeping construction time to a minimum saves the owner money, plenty of it. Example: A building with a construction loan of $150,000 should be completed in seven months, but it takes a month longer. What happens to the loan? The owner loses more than $1,000. Here is why. By the seventh month, probably 80 percent of the construction loan money has been paid out by the lender and costs the owner interest, let's assume 12 percent. 80 percent of $150,000 is **$120.000.** On that portion, the yearly interest is $120,000 x 12% = **$14,400.** Interest for one month is $14,400 ÷ 12 = **$1,200.** Diddling around for one more month in construction costs the owner $1,200. If the building is an income property, he also loses one month rent in addition.

What causes construction to lose momentum?

a. Contractor has too much work, spreads his forces too thin or he does not have his subcontractors under control. Or both.
b. Owner does not make necessary decision on time—change orders, selection of materials—or does not pay the contractor on time!
c. Material shortages, labor disputes, bad weather.

What to do? The answers to the questions above are:

a. If there is not a steady flow of activity on your project, have a meeting with your architect and the general contractor. If you have a watertight contract, the contractor is obligated to provide adequate working forces. There are days when little work can be done, after plastering or drywalling, for instance. But those days are few—fewer than a tardy contractor would like you to believe.

How to get your project on the contractor's front burner again is spelled out in the General Conditions of the Contract for Construction, Form A-201, under "Progress and Completion" and consists, briefly, of the following steps:
• If meeting and talking with the contractor does not improve the work progress, then the owner should send a registered letter to the contractor demanding "that the

contractor carry the work forward expeditiously with adequate forces and shall achieve Substantial Completion within the contract time."

- If the contractor does not resume work "expeditiously" within seven days, the owner should have the architect or an attorney send a second, registered letter threatening termination of the contract.

b. Owner has to get off the dime. Change orders, material decisions should be made within five working days at most. Pass on the contractor's payment request immediately to your architect, lender or fund control.

c. **Material shortages:** Any good contractor can find the material somewhere—in Canada, in New Zealand, in Timbuktu. He can find the stuff if he tries hard enough; it may cost a little extra. He should have known where to get materials when he took on the project.

Labor disputes: Not likely; most small projects are non-union. If it is a union project and they hit you, cancel the contract and hire, as owner/builder, non-union subcontractors. Discuss this situation with your attorney.

Bad weather: Tough luck. Don't start trenching in December if you can avoid it.

DO:
- Insist on adequate working forces throughout the project. Be resolute.
- Discourage work after daylight and on Sundays and holidays.
- Have the building roofed, closed in and locked up as soon as possible.
- Expedite your own decisions on material selection, changes, substitutions.
- Pay contractors, architects, consultants on time—religiously!
- Forward contractor bills to the lender or fund control within five days.
- Have the permit set on the job site. And only for use by inspectors. At other times, hide it! (Not in the job toilet.)
- Protect pedestrians and adjoining property.
- Observe local ordinances regarding construction noise, working hours and work on Sundays and holidays.

DO NOT:
- Keep any outdated drawings on the jobsite.

C-3 THE CONTRACTOR USES MY PHONE, MY ELECTRICITY AND MY BATHROOM!

Construction workers need utilities on the jobsite: a telephone to order materials or to call the home office; water to mix mortar, keep concrete slabs wet, wash ceramic tiles, clean tools and to drink when out of Schlitz; electricity for power saws and nailing guns; fuel gas to test ranges, furnaces and log lighters. A toilet on the jobsite is required by law; it can be the owner's bathroom or a rented portable toilet. There should be an understanding in writing between owner and contractor as to who pays for these utilities.

On remodeling projects, it is easiest if the owner agrees to pay for the use of water, sewer, fuel gas, electricity and local and toll calls by the foreman or subcontractors. For any project that takes longer than two or three weeks, the general contractor should order a portable toilet, at approximately $35 per month, and include the cost in his proposal. There is a space on the contractor's cost breakdown sheet for "utilities and job toilet." It is not worth aggravating the lady of the house by having the ditch diggers drag their clay feet over her snow white sheepskins in the master bedroom.

On new projects the general contractor should pay for all utilities and job toilet until the owner takes over the building.

While the above outlined arrangement is standard procedure, it should be spelled out in writing so the contractor knows which costs to include in his proposal. If it is not specified, and there are several general contractors bidding, some might include the costs, others might not. The place in the contract documents to tell the bidding contractors what is expected is either on the general notes sheet of the drawings or in the special conditions of the specifications. If someone forgets to mention utilities in either place, the arrangement could be spelled out in the agreement between owner and contractor, under "Miscellaneous Provisions." Instructions regarding utilities could read:

Remodeling Project:

Owner will pay for water, sewer charges, fuel gas and electricity used during construction and for local and area toll calls related to the construction project. No personal or long distance calls. General Contractor shall order and pay for portable job toilet, acceptable to the Building Inspector.

New Project:

Contractor shall pay for utility inspection fees, water, sewer charges, fuel gas and electricity during construction and also for installation and use of job telephone and job toilet. Toilet will be used by all workmen including those under separate contracts and is to remain on jobsite until the last subcontractor under the general contractor's supervision has completed his part of the project.

DO:

- Specify who pays for installation and use of utilities, job toilet and telephone.
- On larger projects specify that the contractor furnishes a field office with phone.

Your building project is on schedule: excavations are made, foundations poured, utility trenches backfilled, first floor in place. Soon the contractor expects the first payment. The next chapter explains construction loan disbursements.

C-4 LENDER'S MONEY COMES IN DRAWS

Most building construction is financed by construction loans. A lender can be a private person, an insurance company, a commercial bank, a savings and loan association or a credit union.

How does the construction loan money get from the lender to, let's say, the masonry contractor? The money is disbursed by the bank in portions, payments called "draws." Each draw is a percentage of the entire loan. That portion will be paid by the lender to the general contractor after the completion of a certain building phase, as specified by the lender. The lender's inspector visits the building site and verifies to his bank that the phase is completed. The lender then pays the general contractor, who in turn pays the masonry subcontractor.

A typical disbursement schedule with six draws looks like this:

    20% of the construction loan is disbursed after foundation is completed, including the under-slab or under-floor heating, plumbing and electrical lines.
    25% after framing, rough plumbing, rough electrical, rough heating and roof are completed.
    20% after drywall and exterior plastering or siding.
     5% after cabinets and pullmans are in place, complete with tops.
    20% after the Notice of Completion is filed.
    10% after the expiration of the time limit in which liens for labor and material can be filed, normally 30 days.

Some lenders prefer to disburse the money in five draws; then there will be no cabinet draw and one of the others is 5 percent larger than indicated. We recommend asking for a cabinet draw; otherwise cabinets and tops have to be paid from the draw paid after the Notice of Completion is filed. Most cabinet shops, marble installers, etc., won't and shouldn't have to wait that long.

Some of the finishing trades—painting, flooring, carpeting, for instance—will be paid from the last draw, due 30 or more days after the building is substantially completed. They will holler and scream and call the owner Saturdays and Sundays. The time limit of 30 days serves to allow any contractor, tradesman, laborer or supplier of material who has not been paid to file a claim against the owner. The lender will be alerted and can tell the person who has received the disbursement to pay up or the last payment will not be forthcoming. If no liens are filed in those 30 days, the lender assumes that everyone has been properly paid and he will release the last draw.

There is a method by which the last disbursement can be made sooner. The owner should arrange with the title insurance company for an INDEMNITY AGREEMENT; most companies will add it without extra charge. With an indemnity agreement, the last draw will be released in three or four days after filing of Notice of Completion. To get the last draw as soon as possible these steps must be taken:

1. Notice of Completion must be signed by the owner and recorded by either the lender, the title company or the owner. Normally, the title company or the lender files. The owner should ask for a copy of the recorded notice; cost (approximately $2) to be paid by the owner. Another copy must be sent to the lender or the title company respectively.
2. After the title company receives proof that the Notice of Completion has been filed, the title company will inspect the project to assure that all work has been completed.
3. Then the owner (and sometimes the general contractor) will have to countersign the indemnity agreement which states, in effect, that owner and general contractor will indemnify the title company if there are claims for unpaid labor, material or services. The title company would first have to indemnify the lender, but then turn around and go after the owner and contractor to try to collect from them. It puts the pressure on the owner and contractor to not make payments without receiving a waiver of liens.
4. Title company will then update the title policy and forward it to the lender.
5. Lender will disburse the final draw, either to the general contractor, the owner or fund control.

What is fund control? Let's assume the lending institution has agreed to give you a construction loan of $80,000 and will disburse it in six draws. How does it work? As previously stated, the lender has the right to insist that the owner "have fund control." Remember the golden rule: He who has the gold makes the rules.

With fund control the lender will make disbursements not directly to the owner or general contractor, but to a service firm, called fund control, which receives, controls and disburses construction money. Similar to an escrow department, the fund control firm acts independently, is not an adjunct to the lender, the owner, the general contractor, the architect or the title company. In many states, as in California, it is regulated by the Commissioner of Corporations and the escrow laws.

For the service of controlling funds, the owner pays a fee, about $3.50 to $4.50 per $1000 controlled. In Chapter B-5, we advised inquiring into the rates for this service and taking the lowest bidder. As with title insurance and fire insurance, the **owner** has the right to select the fund control firm, which, however, must be acceptable to the lender.

To get fund control started, inquire about the rates, ask for an agreement form, fill in the blanks, sign it and send it back to the service firm. (See a typical form on the next page.) The entire service charge will be deducted from the first disbursement. On the cost breakdown sheet should be, of course, an allotment for fund control. In addition to controlling funds, the firm can be helpful in a number of small ways: knowing which lenders make loans at a given time, assisting with bonds and lien releases and furnishing credit information on contractors.

FUND CONTROL AGREEMENT

Account No. _____

TO SOLANA FUND CONTROL INC. hereinafter referred to as "Control"

in consideration of your services in handling the funds placed with you for (Construction) (improvement) of a _____

_____and improvements (under contract between) (being built by) _____

_____Owner (as Owner-builder)

and _____Contractor.

on _____property legally described as follows:

_____Lot____Block____Add____

_____ hereinafter referred to as the "job." Control shall receive as compensation for its services hereunder

the sum of $_____ to be deducted from the funds as soon as they are placed in the hands of Control. The sum of

$ _____ will be deposited in accordance with the following schedule:

_____

_____

The undersigned hereby assign to Control and authorize and instruct _____

to disburse to Control any and all proceeds from any construction or improvement loan relating to the job.

You are instructed to disburse said funds only upon the written order of _____
who agrees to order disbursement of funds only in payment of the cost of labor, materials, services, permits, fees, or other items of expense incurred in performing the job and not otherwise. You are authorized to presume any written order executed by the person(s) authorized herein is given in accordance with the terms of this agreement and that funds disbursed pursuant to said order are for the purposes stated therein.

Contractor agrees that he shall not assign any of his rights to funds held under this agreement without the written consent of the Owner and Control.

Any extras or change orders negotiated for by the Owner and the Contractor shall be set forth in writing and shall set forth a stipulated sum for each such extra or change. Any payment for extras or change orders shall be made into the Fund Control prior to the performance of the extra work that may be called for. Any deletions that reduce the contract price to Owner shall be refunded to the Owner on Contractor's written order on Control.

It is understood that these instructions cannot be revoked or modified without your consent.

Should any controversy arise between the undersigned, or any person, you shall not be required to take any action, but may withold all moneys, without interest, or other things deposited with you until such controversy shall be determined by agreement of the parties or by proper legal process.

The undersigned, jointly and severally, agree to hold harmless, indemnify and defend Control from any and all claims, demands, liabilities and suits of every kind and description, including those of the undersigned, made or brought for, or on account of, or alleged to have resulted from Control's administration or disbursement of the funds, or any act of Control, arising out of this agreement, and further agree to pay Control's reasonable attorney's fees and cost incurred with or without suit in connection therewith.

It is further understood and agreed that in accepting this position as Control you are not to be responsible for nor guarantee that the job shall proceed, that the costs thereof shall be paid, or that the job will be performed in accordance with plans and specification, nor have any duty to inspect the job or to determine that any labor and materials used in the job are in accordance with the plans and specifications, or to determine that any funds disbursed are for the purposes stated in the written order directing disbursements.

The Contractor shall have no right, title or interest in any of the funds held under this agreement until Contractor certifies to Control that all cost of labor, materials, services, permits, fees and other items of expense incurred in performing the job have been paid and furnishes to Control an agreement indemnifying Control from any claims arising out of this agreement, including Control's reasonable attorneys' fees and costs incurred with or without suit. However, Contractor shall be entitled to be reimbursed for cost of labor, materials, services, permits, fees and other items of expense incurred in performing the job which he certifies he has paid, and can provide paid receipts acceptable to control.

In the event of the death or legal disability of the Owner or Contractor, or any of them, you may, at your option, continue to act until completion of the construction or return all unexpended funds to the lender and thereby be discharged of any further liability hereunder.

In the event there is no disbursement from the fund control account for a continuous period of ninety days, you may disburse the entire balance thereof, with the consent of the lender, to the Owner and thereby be discharged of any further liability hereunder.

This agreement is not made for nor intended to be for the benefit of anyone not a party to this agreement, including those furnishing labor or materials and any lender, but is made for and intended only for the benefit of the undersigned.

Dated this_____day of_____, 19____, at_____, California.

ACCEPTED

By _____   _____
                          Control   Owner

                                                                                   _____
                                                                                                              Owner

                                                                                   _____
                                                                                                            Contractor

This form reproduced with permission
from: Solana Fund Control

How does the money get from the lender to the masonry subcontractor **under the fund control** system? The masonry subcontractor sends a bill to the general contractor (or owner, if owner/builder project). The general contractor (or owner) decides from which draw the masonry contractor will be paid, sometimes in partial payments from several draws. The general contractor (or owner) sends an "Order on Fund Control" to the masonry man (see sample form below). The masonry sub signs the "Certificate and Waiver of Lien" on back of the order and mails it to fund control. As soon as the respective draw arrives at fund control from the lender, the fund control firm writes a check to the masonry contractor.

A few reflections: Fund Control will not guarantee that the mason's or any other subcontractor's work was completed per contract or (sic) was even done at all, but will rely entirely on the person who writes the Order on Fund Control.

Fund Control cannot disburse monies before they are received from the lender, nor can it pay out more money than was allocated for each trade, vendor, etc., on the cost breakdown sheet. If the subcontractor's Order on Fund Control is more than was allocated, the balance will have to come out of the contingency allowances.

Once the lender has disbursed the entire construction loan to Fund Control and Fund Control has paid contractors, vendors and the owner's reimbursements, all per submitted Orders on Fund Control, that's it. If more bills come in the owner has to deposit additional monies with fund control or pay the contractors directly. If he pays directly, he should get a Waiver of Liens, just as fund control does. (See backside of Order on Fund Control.) Sounds cumbersome. And it is. But remember: it is YOUR money.

JOB NAME

DATE

TO

FOR

BAL.

PAID

BAL.

№ 14896

FORM NO. 736

Order on Fund Control

WESTERN of San Diego FUND CONTROL, Inc.

a Boise Cascade Company
P.O. BOX H   2745 TIDELANDS AVENUE
NATIONAL CITY, CALIFORNIA 92050
PHONE 474-3341

CHECKS ISSUED BETWEEN 1:30 PM TO 4:30 PM ONLY

CHECK NO.

DATE

ITEM NO.

DATE

PAY TO

ADDRESS

THE SUM OF   FOR

FROM FUNDS HELD BY YOU FOR DISBURSEMENT ON JOB NAMED BELOW.

JOB NAME

JOB ADDRESS

FOR THE PURPOSE OF OBTAINING PAYMENT, THE UNDERSIGNED CERTIFIES THAT THE LABOR AND MATERIALS REFERRED TO HEREIN HAVE BEEN ACTUALLY FURNISHED BY THE PERSONS NAMED HEREIN IN THE AMOUNT SPECIFIED AND WERE USED ON THE JOB DESIGNATED HEREIN.

№ 14896

CONTRACTOR

Typical Order on Fund Control
(Courtesy Western of San Diego Fund Control Inc.)
The back side of this form is a Waiver of Liens
and is shown in Chapter C-13.

HOW FUND CONTROL WORKS

Note: Either the owner or the general contractor can contract with Fund Control and then issue the "Orders On Fund Control"

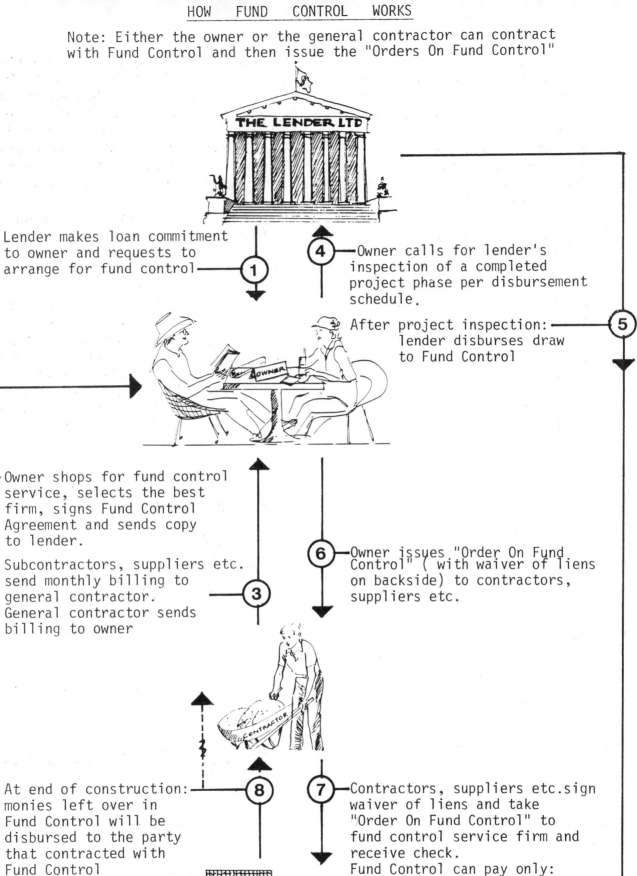

THE LENDER LTD

Lender makes loan commitment to owner and requests to arrange for fund control — (1)

(4) — Owner calls for lender's inspection of a completed project phase per disbursement schedule.

After project inspection: lender disburses draw to Fund Control — (5)

(2) — Owner shops for fund control service, selects the best firm, signs Fund Control Agreement and sends copy to lender.

Subcontractors, suppliers etc. send monthly billing to general contractor. General contractor sends billing to owner — (3)

(6) — Owner issues "Order On Fund Control" (with waiver of liens on backside) to contractors, suppliers etc.

At end of construction: monies left over in Fund Control will be disbursed to the party that contracted with Fund Control — (8)

(7) — Contractors, suppliers etc. sign waiver of liens and take "Order On Fund Control" to fund control service firm and receive check.
Fund Control can pay only:
  a) after lender's inspection
  b) after receipt of draw
  c) per construction cost breakdown

FUND CONTROL

DO:

- Have the first draw as large as possible in order to be reimbursed, as owner, for land payments, survey, architect fees, permits, etc., as long as sums were allocated for these items. You will, however, pay interest on these reimbursements from that moment on. If in doubt, get advice from your lender or architect.
- Arrange for a cabinet draw.
- Arrange for an Indemnity Agreement with your title insurance company **if** you intend that you (or the lender or fund control) will make the final payment to the general contractor **before** the lien period (30 days) has elapsed.
- Shop around for the best rates on fund control. Choose a service firm close to your place of business if you have a choice. You might want to pay a visit and discuss the fund control status.
- Allocate the fund control fee on the construction cost breakdown sheet.

DO NOT:

- Call the lender for a project inspection and disbursement of a draw earlier than necessary since you the owner will pay interest as soon as the draw leaves the lender's office.
- Agree to take the construction loan in a lump sum, as is sometimes done with smaller remodeling projects and owners with excellent credit rating. You probably have no other place to put the loan money than into one of those silly 5¼ percent accounts, while you pay out 11 to 18 percent or more to the lender.
- Agree to pay dutch interest.
- Pay any advance to any contractor for any reason.

We encourage clients not only to be involved in arranging for and disbursing the construction loan but also to make building decisions. C-5 advises how to plunge right in.

C-5 THE SELECT AND CONFIRM LIST

A client who added to her hillside house asked her architect to make up a list she could hang next to her mirror, to remind her of her part of the work during the construction phase. The architect made a sign, about 2 feet by 1 foot; it had three words, with a colorful flowery ornament around those all-encompassing words:

DECISIONS
DECISIONS
DECISIONS

The sign hangs in her den now, as a memento of the exciting months she had to make all those decisions, with the help of her husband, her children, the neighbors, the neighbors' friends, the lumberyard people, the tilesetter and, of course, the architect. Besides the decisions, decisions, decisions plaque she was furnished with three lists during the construction phase. The first one was called "Select and Confirm."

During the planning and bidding phases, a client cannot select every term irrevocably. Let's say you want ceramic tiles on your kitchen counter—a simple example. The drawings call for ceramic tiles, the specifications describe a certain good quality tile. Yet when drawings and specifications are sent out for bid, is your selection really final?

Did you consider laminated plastic in lieu of tiles?

Will the tiles, as specified, be available on time?

Did you have time enough to consider other tiles, visit showrooms, compare prices?

Should the grout be darker, as specified, or maybe lighter, or the same color as the tiles?

The architect advised laying the tiles in mortar, but to save money they could also be glued on.

Drawings and specifications should be as complete as possible at the time of bidding and even more so at the time construction starts. Yet a number of decisions will still be pending. The owner should have a list of these questions: selections still to be made, options to be considered, or just tentative selections to be confirmed. The architect should make up the list for the owner, organized in the sequence of construction. Most clients welcome this list. For a **typical custom residence,** a select and confirm list could look, in part, like this one:

SELECT AND CONFIRM LIST

Dear John and Marge,

Before or during construction, please select or confirm the following items as tabulated below in the sequence of construction. Look up in your drawings and specifications what was tentatively selected; if your ultimate choice is different, let us know. Decisions must be made soon so that materials can be ordered ahead of time; some items need as much as three months. Call us if you have questions.

During Grading:
Where should top soil and excess fill be deposited?

Floor and Framing:
Did you add to your fire or homeowner's insurance?
Living room concrete slab: confirm location of floor outlets for floor lamps.
Did you notify the vacuum system installer? His work is under separate contract.
Order intercom (separate contract).
Review and verify door and window opening locations.
Where do you want tinted glass? Take glass samples out to the site.
Select door and cabinet hardware now.
Order soft water system (separate contract). What type of drain is needed? Inform general contractor.
Verify all towel bar locations for solid backing.

Rough Plumbing:
Verify those bibb locations and points of connection for your future sprinkler system.
Consider installing vertical sleeves in flat garage roof for future solar panel installation and connection lines to water heater. Inform general contractor.

Rough Electrical:
Review all electrical outlets and switches; add or adjust before electrician strings wires.
Verify height of wall light fixture outlets.
Remind electrician of dimmerswitch locations and that you want the true energy-saving dimmerswitches.

Rough Heating:
Review thermostat location (efficiency, decorating).

Drywall:
Confirm locations where you want wallpaper; there will be no spray on those walls. Where do you want concealed drapery tracks?

Plastering:
Confirm exterior of main fireplace: plaster or brick veneer. Brick veneer is an alternate in your contract.

Cabinets:
Verify with cabinetmaker location of all drawers, open shelving, doors, tray storage.

Ceramic Tile:
Do you want grout color to be lighter or darker than tiles? We think: somewhat darker.

Finish Carpentry:
    Select, pick up and deliver to jobsite all kitchen and bathroom accessories as listed under "Building Specialties" of the specifications.
    Where do you want double poles at wardrobes?
    Verify height of wall bench in den; is now 20"; should be 15" if you want to upholster the benches.

Finish Plumbing:
    Inspect and verify all plumbing fixtures and fittings before installation.

And so the list goes on . . .

Note: **The items on this list are not typical** of the average building; they were taken from the select and confirm list of a particular house.

      Labor
+ Materials
+ Expenses
+ Overhead
+ Profit
_____
= Contract Cost

To keep the contract cost low, the cost of every component must be low. Material, for instance. After you have selected **what** you want, buy it as low as you can. Remember the old Broadway play, "OF COURSE, YOU GET IT WHOLESALE!"
Hey, that's our next chapter.

C-6 OF COURSE, YOU GET IT WHOLESALE!

An American Standard bathtub, "Bildor"
5'-0" enameled cast-iron, white,
slip resistant, retails for $338.00
The plumbing contractor pays .... <u>230.00</u>
He makes a profit of $108.00

Should you try to buy the bathtub wholesale yourself, pocket the $108 and just let the plumber install it? Same with lavatories, disposers—maybe everything: lumber, gypsum board, nails, roof shingles—you buy it wholesale, then just hire the bodies to nail your building together? Would you save a bundle, namely the difference between wholesale and retail, say about 30 percent? Probably not.

The message: Don't try it. For a number of reasons:

► Contractors cannot charge a 30 percent markup; competitive bidding prevents that. If plumbers, carpenters, electricians or tilesetters charged retail prices for the materials they use, they would seldom get the contract. Of course, if you build under "time and material" instead of competitive bidding, you could pay $338 for the tub. If a particular contractor is eager to get the job, some markups are as low as 6 or 8 percent. But the trades need a reasonable markup on materials and equipment to pay for overhead (getting material quotations, pickup, storing, delivery to the jobsite, guarantees) and for profit. Most contractors would not submit a bid for installation only of owner-furnished materials.

If the total cost of owner-furnished material plus contractor's installation were compared to contractor-furnished material including installation, the owner would find that in most cases he is better off to let the contractor furnish materials and install. Only very experienced owner/builders should buy materials directly and hire out the installation, but certainly not plumbing, electrical or heating work.

► Most owners wouldn't know where or how to buy; and if they knew they probably wouldn't get the trade discounts. Wholesale houses don't like to sell directly to the consumer/owner. One reason is not to anger the tradesmen. Another is that tradesmen know what they want; transactions are quick and to the point. But to let any Wishy B. Washy roam for hours through bathtub catalogs and advise her and explain and eventually sell her one — the fellow at the wholesale counter would not have enough Rolaids.

Upshot: Under tight competitive bidding, with precise drawings and specifications, the owner automatically buys most materials as close to wholesale as is possible.

There is another good reason to let the contractor buy most materials: he will use them more carefully, more economically. Construction lumber for instance: if the contractor pays for it he will not allow a carpenter to cut a 14-inch long blocking piece of 2 x 4 from an eight-foot-long stick taken from the stud pile.

And yet, we advise clients of smaller projects (residential remodelings, new residences, commercial or church expansions, etc.), where the owner takes a more personal interest in the building, to buy certain materials or equipment directly. Not because the owner can buy for less than through the contractor, but because it gives him more control over the type, appearance, quality, delivery date or all these criteria.

Such items include: light fixtures, low voltage equipment, bathroom and kitchen accessories, drapes and drapery tracks, carpeting, special ceramic tiles, built-in radios, intercoms, garage door openers, whirlpools, barbeques, topsoil, planting materials, sprinkler systems.

Next question: How, then, to buy these and other building materials and equipment at wholesale—or close to it—when you, the owner, buy directly? It is easy, if there is a contractor involved with your project. (And, remember, for most jobs we recommended that.) There are two methods:

1. Your contractor (general or sub-) establishes an open or "tab" account with the supply house from which you want to buy; with many suppliers he probably has such an account. ("Tab" refers to a plastic card the supplier keeps with the contractor's name, number, etc.) Most tab accounts are established not only in the name of the contractor but also in the name of your project address; it facilitates a mechanic's lien, if one ever has to be filed. Have your contractor add your name to the account; then you go to the supply house, make your selection, sign the invoice and later reimburse the contractor. Or the contractor can add the item to your monthly billing. (We assume the item is not part of his contract.) This means that as long as you have a contractual relationship with a contractor you can buy almost any building material or piece of equipment at contractor's prices. We see no reason why a tile contractor should not only let you buy specially selected tiles on his accounts, but also a doormat, a whirlpool or light fixtures. If the item is part of the contract, then you really only select and pick it up. Just sign the tab account invoice.

2. If the contractor does not have an account at the kind of supply house you need and does not want to open one—or cannot because he lacks the credit rating at that time—here is a little detour by which you can still buy at contractor's cost. Sounds cumbersome, but it works: The contractor phones the supply house that he needs whatever-it-is and that you, the owner, will come by and select it for him. Then he gives you one of his checks, blank, with his contractor's imprint on it. You go, select or order the item and pay with his check. The wholesale supply house sold to a contractor, right?

Back home you reimburse the contractor. If you have time to think it through, this method could have some tax advantages to both you and the contractor. If you have qualms, don't think it through.

Most contractors welcome an owner selecting and delivering to the jobsite certain materials, built-ins or equipment because it saves them time and often frustrations. Nothing drives a contractor bananas as much as traveling from supply houses to your kitchen table, back and forth, and spreading out before you and your friends and neighbors tile samples and door knob samples and carpet samples and paint chips; oven catalogs and sprinkler head specimens. And all those exotic brochures with light fixtures: Baroque, Colonial Williamsburg, Avant garde, Altdeutsch, Spanish. Ask any contractor—he would rather be sailing!

Some fine points: Whatever you buy, as long as it is budgeted, will be paid out of the construction loan, either from the allotments for the particular trades (light fixtures under electrical work), from "miscellaneous" or from "contingencies." If you need a barbeque or a whirlpool or a low voltage set, it is smarter to buy with construction loan money and pay it off over 25 years at 11 percent than to buy it with a bank credit card or at the downtown department store with even higher interest rates. Again, assuming you decided not to buy cash and carry.

Be careful not to buy with construction loan money anything that will not be installed or used on the project for which the loan was made (for instance on your brother's patio or in your girlfriend's townhouse). Besides probably not having that extra built-in whirlpool or barbeque in the budget, legally "any person who submits a false voucher to obtain Construction Loan Funds and does not use the funds for the purpose for which the claim was submitted is guilty of embezzlement." (Penal Code Sec. 484C)

More: Any person who uses material or equipment meant to be installed on a commercial or income project but installs it at his private residence becomes guilty of tax fraud.

DO:
- Leave most material and equipment buying to your contractors.
- Get contractors' discounts on everything you buy directly, using:
  a) contractors' open or tab accounts
  b) check interchanging
  c) your architect's resale license.

DO NOT:
- Buy with allocated construction loan money anything that will not be installed or used at the project for which the loan was obtained.

Construction is underway, the contractor receives his first payment, you select materials. Sometimes you wish there was a pause to review the drawings, to understand better what looks like chaos to you but beautiful progress to the contractor. Who will supervise all this progress? Read on.

C-7 WHO SUPERVISES WHOM? WHY? WHEN? AND HOW?

In a single month on different projects under construction these mistakes were observed:
- The wrong type of nail was used on a tile roof.
- A new residence was built with a 76-foot front setback instead of the 85 feet required (all foundations were formed!).
- The footing reinforcing of a new addition was not doweled into the existing footings.
- Finish grades did not slope away from a new school building.
- A cabinetmaker forgot to provide a space for the trash compactor in the kitchen cabinets.
- The wrong forced-air furnace was installed.
- A 220-volt electric washer outlet was not grounded. (This could be fatal.)

And there were more.

The contractors were lousy, you might say. No. They were good contractors. Mistakes happen during construction; some are small goofs, some are misinterpretations of the drawings, some are blatantly irresponsible errors. Without elaborating on who was at fault and who should be liable for these mistakes, the fact remains that for almost all mistakes the owner pays—in money, in construction delay (which is money), in inconvenience, in inferior equipment or in an inferior solution to a building problem. The owner pays but he is not always aware of it.

The message: Building construction must be tightly "supervised," from the day the layout man stakes the front setback to the day the painters touch up the carpetlayers' dirty spots on the living room walls.

"Supervised" we put in quotation marks for a reason. The word "supervision" is almost a legal term and should not be used loosely; at least the client should know who supervises and who does not; and why, and when and how. We are not going to get bogged down in legalistic semantics, but let's use the correct terms and understand what they mean to the owner, and how important they are to his satisfaction and his pocketbook.

The general contractor supervises. He orders materials and directs his personnel and subcontractors; he has control of the work and is responsible to the owner for the performance of those under his control.

The owner and/or architect observes. He will have to decide during those four to 10 months of construction to accept or not accept what the contractor and his personnel produce. And, the owner should know he can only expect what was agreed on in the contract documents—the drawings, conditions, specifications, addenda, change orders and the agreement.

To accept what the contractor produced! Not only on large projects, no—on most building projects, down to the add-a-bath-and-a-bedroom remodeling. Few owners have the time and the knowledge to judge the contractor's performance. Quite a few of our clients thought they had the knowledge, at least the determination, to deal with a contractor. Most of them didn't. In nine out of ten cases, to safeguard his interests, the owner is well advised to hire someone to observe critically but in a positive manner the contractor's performance.

The most logical choice for overseer: the architect or engineer who designed the building and produced the contract documents. That person will "observe" (that is the legal term now), and report to the owner. The contractor supervises, therefore "supervision" is not part of the duties of the architect. The correct term for the full activity of the architect during the construction phase is "administration of the construction contract." This includes:
* Interpreting contract documents; writing change orders
* Establishing standards of acceptability
* Checking shop drawings and contractors' submittals
* Judging performance of contractor; making inspections
* Issuing certificates to make payments to contractor
* Assisting owner in making selections or changes
* Determining date of substantial completion; filing Notice of Completion
* Writing punch lists
* Making guarantee inspection one year later

Why is this important, this checking up?
▶ To insure that the contractor delivers what he has contracted for: a building of a certain size, with a certain arrangement and specified quality level, which includes types of materials and workmanship.
▶ To insure the delivery of the completed project on a promised date.
▶ To spot building mistakes as well as necessary or desirable deviations from the drawings during construction.
▶ To keep the momentum of speedy construction.

When to do all this? On larger projects the architect should be hired for the complete administration of the construction contract. The cost is minimal, considering what a capable architect can save the owner. The full service is around 1½ to 2½ percent of the construction cost—a bargain. For a $120,000 office building, that's approximately $2400 spread over six to eight months for the work described above. Even less for residential work, as you will see. (Insider's note: Most architects do not make a profit on this phase of the project but will perform the task conscientiously.)

What can the owner do in the construction phase to help himself and his project, besides listening to the esoteric dialogues between contractor and architect and then paying the bills at the end of the month? A darn good question. On smaller projects, especially residences and remodelings, the interesting task of construction check-up can be shared between the owner and the architect, if the owner is intelligent, eager and interested in saving some of the fee. Here is one good method:

Have the architect stand by for the periodic visits to the construction site and conferences with the owner. For building a residence or residential addition, the minimum

166

job visits would be:

  After stake-out
  Before footings are poured
  After footings and slab or floor are installed
  Twice during framing and roofing
  Before drywall
  Twice during cabinets, plastering, floors
  During painting
  Pre-final, to write the punch list
  Probably a final final!

Say 11 visits at 2 hours each . . . . 22 hours
      10 office meetings . . . . 25 hours
      Total . . . . 47 hours

At $32.00 per hour, that's $1504 or about 1½ percent of a $100,000 residential project.

To help clients understand the construction process so they can really observe and check the contractor as well as cooperate with him, we issue a list of items the owner should inspect and check off. The tabulation is in order of the sequence of construction. It is called the "client's construction checklist," the architect's second list to the client.

Clients are urged to use the list and call the architect with questions. In order to alleviate the client's hesitation to call, even with minor questions, our office does not charge for telephone conferences with clients.

Socrates assured us that the "unexamined life is not worth living." In regard to your building project, we assure you that the unexamined construction project is not worth paying for.

DO:
- Have your building checked during the construction phase by a professional other than the contractor.
- Consider hiring an architect on an hourly basis, with an hourly rate of compensation.
- Opt for full architect's services if you don't have the experience or the time to "administrate" your contract competently.
- Ask your architect to prepare for you a client's construction checklist if you are an owner/builder or are otherwise interested in overseeing your project. More on this in the next chapter.

C-8 CLIENT'S CONSTRUCTION CHECKLIST

The list runs three to six pages and is given to the client when construction starts. The following is the very first part of such a list. The items mentioned are not typical for every building; typical only is the concise tabulation, in the sequence of construction.

Project: Bentley Residence
Construction Administration
Page 1 of 3

John and Marge,

The select and confirm list you have received previously dealt with selections still to be made or confirmed, or alternates to be considered. This construction checklist deals with correct installation and good workmanship, and is an extract from the "Installation" and "Workmanship" sections of the specifications.

In General
Give a copy of this list to the general contractor.
Obtain from contractor certificates of insurance.
No outdated drawings on building site.
Job toilet to be installed before construction starts.
Have handy: telephone numbers of nearest hospital and physician.
Call us before giving instructions to general contractor.
Never give instructions to subcontractors or workers.
Change orders to be in writing, unless the change does not alter the cost or is below $25.
If the work pleases you, send out a pot of coffee or a six-pack once in a while.
Call us if you have any doubts or questions; don't wait until our next inspection visit.

Slab and Framing
Consider salvaging good topsoil; designate an area.
Tell the general contractor where you want scrap lumber piled; you might want it for
    firewood later.
Before pouring slab: all wood chips and sawdust to be removed as much as possible.
No portion of slab to rest on any live vegetation or loose fill.
Check minimum depth of footings; see drawings.

Slab thickness 4 inches net at the house; 3½ inches at the garage.

Vapor barrier not to be pierced by reinforcing or tools; overlay pierced holes with another piece of barrier.

Along certain walls, separate slab from stem walls. See foundation drawings.

All wall pilasters to rest on footings; no pilasters cantilevered from walls.

Forming lumber to be removed completely before backfilling; no broken off stakes left in ground.

End of sample. The list goes on with masonry, carpentry, roofing, windows, doors, all the way to carpeting and touch-up painting.

DO:

- Ask your architect to furnish you a construction checklist if you are interested in helping to oversee the construction.
- In overseeing the construction: look for the important. Don't nitpick. It is normal that reinforcing bars are rusty, that concrete floors have hairline cracks.

    (Construction) wisdom is partly the art of knowing what to overlook.
- Understand that no building is built the way it is drawn up or specified. If changes are necessary, make them.

Building inspectors can be a real help in overseeing construction. Often there is more than one inspector. You will see in the next chapter.

170

C-9 CHERISH YOUR INSPECTORS—ALL OF THEM!

Your dream house was designed by a registered architect, checked by a city engineer, stamped all over by the Building Inspection Department and, during the bidding period, reviewed by experienced licensed contractors. None of these fellows told you what you now hear from the field building inspector after the foundation trenches have been dug. On the rear side of your house the trenches have to be 12 inches deeper and additional reinforcing bars must be installed in the fireplace footing.

Why? The inspector says the soil is mushy and should be removed by those 12 inches. Something wrong with the drawings? Probably not. Soil conditions can cause a problem during construction, even if there has been a soil investigation. So the inspector finds a solution—deeper footings, more reinforcing, etc., and you pay for the additional work.

The message: During construction changes will be made, not only by the owner, the contractor or the architect, but also by the building inspector. And extra work caused by the inspector's decisions may cost additional money or mean delay in the work. Whether the architect, the city engineer and the contractor overlooked or did not anticipate some item, or mistakenly thought they had interpreted the building code correctly, the fact remains that the field inspector has the right to demand corrections or changes. Normally such requests will be sustained by the Building Department. The Uniform Building Code (UBC), Section 302, states:

> The issuance of a permit based upon drawings and specifications shall not prevent the Building Official from thereafter requiring the correction of errors in said drawings or specifications or from preventing building operations being carried on thereunder, when in violation of this code or any other ordinance of the city.

As said before, the UBC deals mostly with aspects of health, safety and the state's energy-saving requirements; therefore changes requested by an inspector serve, in most cases, the interest of the owner. It would be nice if the UBC and the inspectors would also be concerned with other matters such as quality of materials beyond the minimum code requirements, workmanship and appearance, but this is the owner's domain.

The inspector might require deeper or wider footings, more reinforcing, even a soil report; he might demand additional bracing of walls, timber connectors, thicker plywood, steeper slopes in plumbing lines; he can request a recheck of drawings or truss calculations—all in the interest of safety and health. He will probably find no fault with sloppily hung doors or unsafe door locks, incorrectly nailed siding or skimpy roof overhangs, with a noisy furnace or cheap flooring. He certainly will not interfere if your wall-to-wall picture window looks out to the neighbor's garbage cans instead of to the distant mountains in the morning mist. (Did you buy a stockplan and not change the window location?) These are not his official concerns. The window, however, must have a proper header (safety), not too much glass area or it must be double glass (energy savings) and enough openable area for ventilation (health). This is the inspector's domain.

And yet if you ask questions beyond those subjects the inspector deals with officially you will receive good answers. If you ask, and he has the time, he will tell you how the entrance door should be hung; that galvanized or aluminum nails should be used with exterior siding; he might even remember a good flooring contractor who worked for other builders in the area. Draw on the inspector's experience as much as you can. Offer him a coffee—*after* the inspection.

The value of a building permit to the owner is in the inspections the permit prescribes for the major phases of construction. Normally for residential work these are:

- Soil and foundation, including reinforcing before concrete is poured
- Electric service, framing, heating and plumbing work, roof construction
- Drywall or plaster, exterior siding
- Final inspection.

Items not passed during one inspection must be kept open and reworked until reinspected. If another reinspection is necessary, there may be a charge.

Some structures, especially commercial buildings, require special inspections: masonry walls, high density concrete, steel framing, field welding and others. Arrangements for special inspections are enumerated on the inspection cards when your building permit is issued.

To the request that he tell everything he sees out of whack with your dream house under construction, the inspector might say: "Yeah, owners want to know or don't want to know, depending on who has to correct the mistake. If it is the contractor, the owner wants to know and wants it corrected, whatever it is. But if it is an owner/builder and he has to correct it, then it's different—he would rather see us going than coming." Still we have found that most inspectors are a ready source of information and that their advice and instructions have a protection value much higher than the permit cost; their service is a bargain. Cherish them, if only for those five or six months of construction!

More good news? There might be more than one inspector!

The lender's inspector will come whenever you or the contractor apply for a draw, that partial payment from the lender to you, to the contractor, or to fund control. (See Chapter C-4.) Of him you will not see much; his only concern is to make sure the prerequisites of that particular draw are fulfilled. He might quickly look over the city's inspection record nailed to the garage door post and then speed on to the next project.

Points of advice: If deviations from drawings and specifications are contemplated or are already made, phone him a day ahead of his scheduled inspection. He has to verify that your building is what he sees on his drawings—the ones you submitted with your building loan request to the lender many months ago. Remember? We also recommend that you, the architect or the contractor, call him the day following the inspection, before 8:30 a.m., to verify that he approved the next draw and when it will be transferred.

If your drawings needed approval from an architectural review board, the board will send out its own inspector, normally at three distinct junctures of your building adventure:

- After the building is staked out—to check all required setbacks, minimum or maximum building sizes, etc.
- During framing—to verify the general shape, height, garage location, roof type, etc., you promised to build.
- Final inspection.

If you build under F.H.A. or V.A. regulations, you will have their inspectors poking around too. Get either one of these booklets: **Building with F.H.A.,** from the U.S. Department of Housing and Development or your local F.H.A. office; or **Building with V.A.,** from the U.S. Veterans Administration. For addresses, see D-4. Needless to say, if you have a copy already, be sure it is the latest!

Special inspectors, hired by the owner, are sometimes required for special construction: field welding, high density concrete, elevators, retaining wall grouting, soil compaction. See A-9.

How to deal with inspectors:

DO:
- Realize their experience can be of great help to you; tap that resource.
- If you have major changes in mind, it is better to tell the inspector before you make the change. Sometimes a sketch outlining the change is all that is needed; the inspector may even take it back to the Building Department to have it approved and stamped. Or you drive downtown to obtain approval, then make the change. Tell him about smaller changes made when he arrives, not when he stumbles onto them.
- Always have the stamped permit set and inspection cards available when he comes. At other times: hide it, or it will be swiped immediately.
- Realize that he may order changes that could mean additional work and cost—and a reinspection.
- Be sure work that has to be inspected again is ready at the next inspection. Otherwise there could be a reinspection fee.
- Do not cover any work to be inspected until the inspector allows you to proceed.
- Take our word: Most inspectors are much more useful than we like to give them credit for. It literally pays to have good relations with inspectors.
- Put the coffeepot on when he drives up. He might appreciate a coffee break—after the inspection.

During construction, relatively few changes
are caused by inspectors, architects or
contractors. Owners make most changes!
Watch it. Slippery slopes ahead.
Read C-10. Twice.

The roof is up . . .

C-10 DO YOU WANT TO CHANGE IT?

You might have heard the story about the young couple who bought a subdivision house still under construction. They asked for one change: leave out one of the three windows in the master bedroom. Six weeks later they received a bill from the developer:

Delete one window $100

They called the office: there must have been some misunderstanding; they did not add a window, they deleted one; there should be a credit! There is no credit, they were informed politely. To process the paperwork to delete a window during construction costs the developer $150. Credit for the deleted window is $50. Net due . . . $100. They couldn't believe it. But it's true.

The story may be exaggerated, but makes a valid point: To change anything during construction costs more than you think. Already we have heard about changes by the contractor, by the architect, by the inspector. The client makes changes, too. Every client does.

Hot tip! Make as few changes as possible. Get the best drawings and specifications you can muster. Have most problems solved on paper before starting construction; avoid changes in the field, especially structural changes when the structure has already been built.

Textbook example: To change a window

a. If a window or door is deleted before the respective wall is erected we agree with the couple—that is a small change, there should be a credit, say $40.
b. If the window opening is already framed but you want the opening smaller, it shouldn't cost much either. Just add two studs each side and install a smaller window.
c. To change a window from one location to another or make it larger costs more; the carpenter has to remove studs, install a large header, order a larger window. Cost: maybe $100 or $150.
d. To change a window opening into a pair of French doors is not difficult; just remove the wood studs below the windowsill, a carpenter's helper can do that. But then the builder has to buy the doors, and French doors are expensive. Also add for hardware,

oak or metal sill and painting. The change sounds easy, but material cost is considerable, say $350. And one window is left over.

e. One client wanted a wider window, in a different location and higher, much higher—one and a half stories high. Walls, roof, exterior siding, interior drywall—all were in place, in a one-story entry hall! It would have been better if he or the architect had visualized the dramatic entry hall in the design stage. We had to take off the roof, raise it by half a story, and so on. Did that cost, boy! Fortunately he owned one of the largest Tequila export company in Mexico.

If you don't own a Tequila company, avoid large structural changes unless, of course, you feel they are absolutely necessary. Major changes also break the momentum of the construction pace. They involve additional costs. To the hard cost the contractor will also add 10 to 16 percent overhead and profit. If the architect has to make new drawings —another additional cost. Sometimes even an additional plan check fee by the Building Department.

Another drawback with changes: If you add work to an existing contract, there will be no competitive bidding. You depend on the contractor already on board. Therefore, ask him for an itemized cost breakdown of the added work and try long and hard to understand every cost item. When you delete work you rarely get proper credit. At least, and rightfully, the contractor will keep the portion which previously was his overhead and profit.

Be sure the other trades follow through and that the work required of them is reflected in the change order—and in the cost! When a window becomes a French door, the electrician has to install an outside light, with switch. The painter has to prime and paint the door with three coats. Door hardware is needed as well as an oak or color-anodized sill.

Again, make the change if you consider it necessary. Otherwise the lesser solution will stand there for the next 40 years. Making the change at some later date would cost more.

No matter who initiates the change—the owner, the architect or the contractor—if it costs more than, say $25, have it in writing. If the cost is less, don't insist on a written agreement; contractors know that small adjustments here and there cancel each other out. He may or may not charge you later on.

Architects use a special form to write change orders; a sample is shown on the next page. The form records the changes to be made, the cost of all trades involved, the original contract sum, how the original sum has been adjusted by previous change orders, the new contract sum, including the last change order, and whether the contract time will be increased, decreased or remain the same. The change order will be signed by the owner, the architect and the contractor, and, of course, everyone will get a blooming copy. To keep paperwork to a minimum, several changes can be combined in one change order as long as the entries are made right there, when the change is ordered. Remember, every signed change order becomes a part of your contract. Never decide to discuss and write up changes at the end of the project.

CHANGE ORDER

DATE OF ISSUANCE:_____   PROJECT:_____

TO:(CONTRACTOR)   OWNER:

                                           CONTRACT DATE: _____

YOUR ARE AUTHORIZED TO MAKE THE FOLLOWING CHANGES:

DRAWINGS ISSUED:

CONTRACT SUMMARY:

ORIGINAL CONTRACT SUM .... $_____

PREVIOUS CHANGE ORDERS THROUGH NO ____ ..... $_____

CONTRACT SUM PRIOR TO THIS CHANGE ORDER ..... $_____

THIS CHANGE ORDER WILL ADD - DEDUCT .... $_____

CONTRACT SUM INCLUDING THIS CHANGE ORDER.... $_____

THE PROJECT COMPLETION DATE AS OF THE DATE OF THIS C.O. _____

ARCHITECT   CONTRACTORS   OWNERS APPROVAL
WERNER.R.HASHAGEN & ASSOC. ACCEPTANCE

BY:_____ BY:_____ BY:_____

DATE: _____ DATE: _____ DATE: _____

OWNER:
ARCHITECT:
CONTRACTOR:
FIELD:

If the change would consist of work that could be easily added at a later date, like laying bricks in a patio or installing storage shelving in the garage, consider having the work done on weekends or after the building is completed, by an individual workman. The contractor may recommend one of his workmen. This way the contractor does not have the additional paperwork, the extra job does not interfere with the contractor's schedule, and the owner saves the contractor's overhead and profit. Think about insurance coverage for that workman—will the contractor cover him? Some do.

DO:
- Anticipate changes.
- Avoid major changes where work is already completed, especially structural changes.
- Write a formal change order when the adjustment of all trades is more than about $25 per change. This, of course, is not a legal or fixed limit. Owner, architect and contractor must sign every change order.
- Consider having the contractor make changes on "time and material" instead of a lump sum proposal. To let a contractor work time and material, however, requires experienced judgment by the owner or his architect.
- For items that can easily be added later consider hiring an individual workman.
- If you change glass areas in exterior doors and in windows, check your maximum allowable glass area. You might have to use double glass in some doors or windows.

C-11 PREDICTABLE PASSAGES AND OTHER CRISES

When additions to existing residences are planned, the homeowner is encouraged to lay out the proposed new area according to the drawings, with garden hoses, broomsticks and old boards. This way he/she gets an idea of the size, shape and interior arrangement of the enlargement. Laid out on the lawn, with hoses and boards, the future building addition at first looks rather large; it seems to take away so much from the existing garden or yard. On the other hand, the individual rooms, also indicated by sticks and boards, seem small. That is an illusion because the third dimension, height, is still missing.

Later, after the slab is poured, the gray expanse of concrete looks smaller than the previous layout on the green lawn. Strange! When studwalls go up, rooms seem to grow and have, of course, too much light. Yet from a distance, say from the neighbor's backyard, the edifice looks overpowering, especially if there is more than one story added—too many sticks!

All clients have to live through these passages, and the more sensitive the client is, the more doubtful glances the architect gets on his appointed rounds. The crisis literally darkens when roof sheathing, roofing and the first layer of the exterior wall finish, black waterproof paper, are installed: the scene virtually blackens. The light that makes it through window and door openings seems ungracefully absorbed by dark surfaces—the waterproof felt, the plywood sheathing of floor and roof.

Here the owner may become really disturbed, and neighbors agree: The new place looks dinky and plumb awful. And for such a long time! Because at this stage, between framing and drywall work, nothing seems to move. Electricians, heating men, plumbers and inspectors forever poke around in studwalls, crevices and attics. Is anything being accomplished? Small relief when gypsum board is installed on walls and ceilings. Rooms look grayer and messier yet. Speaking of messiness, wait until the drywall tapers and sprayers move in. We also can predict: They all have dogs and loud radios, at least in California.

Should you call your psychiatrist? The architect tries to sound reassuring: "When everything is painted Navajo white it will all be beautiful."

Hausfrau: "When will that be?"

Architect: "Soon. I swear by my straightedge."

She wishes she had never started building. "Will the whole building be so difficult?" She prays for a good answer.

Well, sort of. A few smaller crises, all predictable, as sure as the sun rises tomorrow. The odds? 999 to 1. For instance (your defense is in parentheses):

- Tilesetters, plumbers, painters will stand with their boots on brand new tubs and toilets. What to do? (YOU cover tubs and whirlpools with paper, tape and boards, as though you want to mummify them for the next 2000 years.)
- The drywall crew will spray over everything—windows, sliding doors, tubs, stepladders, garage floor, your cat. (Tell the contractor you want all overspray removed before final payment. That will perk up his ears.)
- Certain plumbing fixtures never arrive on time. (Plumber should order fixtures the same week he receives his contract. Same goes for electrical light fixtures.)
- At least three items go haywire with electrical installation: half-hot outlets—which lights should be on dimmers—location of special outlets. (Check drawings; check electrical rough-in; check system before drywall is installed; recheck fixture schedule; check, check, check.)
- Painter never has enough money in his contract to do an outstanding job, let alone to shellac knots and paint tops and bottoms of doors. And so much of the appearance of a building depends on painting work. The painter can louse up a good building; or he can save a mediocre one. (Have painting work bid separately from general contract.)

Are these eerie passages normal? Yes, they happen on most projects, from residential work to church buildings to skyscrapers. Will you survive and ever become happy again? Yes, with a good contractor, precise construction documents and thorough inspections. These ups and downs are challenges to the owner and to the architect. And to the contractor, as long as you let him know about your qualms.

Where is the turning point? When the painters and the carpet people move in, when the light fixtures are installed, the drapes and the dimmer switches, the rooms become brighter and brighter, some even appear cozy. The heavy ceiling beams, dropped soffits, upholstered windowseats, subtle skylights—when all these blend together beautifully, that is the turning point.

Your second building project will be a cinch. Crises? Shucks! You can even predict them!

Statistics show that of every 10 people who seriously think of "getting something built" six will eventually build. Out of the six only three will be satisfied with what they get—certainly a dismal ratio. If a contractor does not perform to expectations, that is a real crisis. We hope it does not happen to you. If it does, the next chapter, and a cognac or two, may help.

Needless to say there are incompetent contractors,
just as there are incompetent members in any
profession or vocation. If drastic action
appears necessary, you should know . . .

C-12 HOW TO DISMISS AN INCOMPETENT CONTRACTOR

Fire him! Tell him to move out—with his tools and fools—never to come back. And you will not pay him for his lousy work!

Can you? Yes.

Does it work? Probably not.

We assume you have a written contract with the contractor. The language of that contract will determine how you can terminate it. All contracts we write for our clients make the A.I.A. General Conditions a part of the contract to which the owner and the contractor (normally a *general* contractor) are the principal parties. According to those conditions, either party can terminate the contract. Sometimes it is the contractor who complains: the owner should pay bills sooner, make decisions faster, interfere less, buy more beer—whatever. But contractors seldom terminate contracts; in most cases it is the owner who wants to get rid of a contractor. Whoever thinks of termination, it is best if both parties try—try very hard—to complete the contract. Breaking up a construction agreement can become a nightmare and, in many cases, does not end the owner's obligation to the contractor.

But if the owner is convinced that the contractor has not performed satisfactorily and has doubt that the contractor will complete the contract in an orderly fashion, these are the steps to take:

1. Make a list of the work which, in the opinion of the architect or you the owner, has not been done per contract documents (drawings, conditions, specifications, addenda, change orders). Send a copy of this list with a letter to the contractor by certified mail with return receipt requested. In the letter state that the contractor has seven days to correct the work and to proceed speedily with the contract or you will consider the contract terminated.

2. Wait seven days from the day you receive the postal return receipt with the contrac-

tor's signature. If the contractor makes an honest effort to correct mistakes, and works diligently, we again would urge you not to terminate him.

The contractor may not agree with every item you have on your complaint list. The architect or a construction expert, even the building inspector, should advise you as to which grievances are reasonable. Don't expect to get something if it is not part of your contract, that is, if it is not shown on any drawing or in the specifications or was not properly added by an addendum, change order or the contractor's specific promise.

We have mentioned in previous chapters that it is difficult to specify neatness, good workmanship and resourcefulness. If you hire a contractor who has none of these traits, he will not acquire them on your job and no court or arbitration board can force him to deliver them to you.

3. If the contractor does not respond, either:

File a Consumer Complaint (see sample) with your nearest Contractors State License Boad. The local representative might inspect your project, meet informally with you and the contractor and offer a solution.

or

The contractor or you request that the dispute be **referred to arbitration.** Get a pamphlet from the local chapter of the American Arbitration Association; the address is in the telephone book. The arbitration board is a panel of appointed construction experts who will examine your case and render a decision. Most construction contracts make arbitration mandatory. Arbitration costs less than litigation and is faster. It costs, nevertheless, and the cost will be split, normally, between the feuding parties. During arbitration procedures, the contractor may continue to work on your project if you both agree that he can do so. Again, to understand the process, ask for the Arbitration Association's literature.

or

The contractor will not come back and may or may not respond to your letter. Your architect will know how to handle the situation in accordance with the General Conditions, Article 14, "Termination of the Contract." If you don't have an architect, or if your architect advises you so, contact an attorney experienced in construction matters. In the meantime, do not make any payments to any contractor, workman or supplier. Do not hire another contractor at this stage to finish the work. If someone filed a mechanic's lien, discuss the matter with your architect or attorney; don't panic.

DEPARTMENT OF

CONTRACTORS' STATE LICENSE BOARD

COMPLAINT IDENT.				CLASSIFICATION				P	DATE		Z
R	D	F		CODE	TYPE	TYPE		R	RECV'D		
G	T	Y	NUMBER	VIOLATION	CONT	INV		Y	MO	DAY	ORIGIN

CONSUMER COMPLAINT FORM

1. Please fill in this form as accurately as possible; giving pertinent dates, monies paid, balances owed, amounts claimed by third parties, if any, and who such third parties are.

2. Please make this report legible and understandable. **Print or type.**

PLEASE RETURN THIS FORM IN DUPLICATE

DATE CLOSED | MO | DAY | YR
(Screening)

INITIALS

DEPUTY

MO | DAY | YR

Date assigned:

---

1. PERSON MAKING COMPLAINT...

NAME:

ADDRESS: STREET AND NUMBER

CITY:

NEAREST CROSSROAD:

PHONE(S): RESIDENCE:
                BUSINESS:

2. COMPLAINT AGAINST...

License No.

NAME:

COMPANY:

ADDRESS: STREET AND NUMBER

CITY:

PHONE(S):

---

3. PROJECT INFORMATION...

JOB ADDRESS: STREET AND NUMBER

CITY: PHONE:

OWNER OF JOB:

ADDRESS: PHONE:

CONTRACT DATE: AMOUNT: $

TYPE OF CONTRACT
☐ NONE   ☐ ORAL   ☐ WRITTEN

AMOUNT PAID ON CONTRACT: DATE WORK STARTED: DATE WORK CEASED: MO | DAY | YR

BRIEF DESCRIPTION OF CONTRACT OR AGREEMENT:

---

4. BRIEFLY LIST ITEMS OF COMPLAINT BY NUMBER...

No.	NATURE OF COMPLAINT

5. HAVE YOU DISCUSSED THESE ITEMS WITH YOUR CONTRACTOR?

YES ☐   NO ☐

Signature_____   Date_____

FOR AGENCY USE ONLY

RESPONDENT: LICENSE NO.:   CLASS:

ADDRESS OF RECORD: STREET AND NUMBER   CITY: PHONE:

OTHER ADDRESS:

PERSONNEL:

Form 131-15 (REV. 1-77)

If it seems that termination of the contract is the inevitable solution, the next steps should be taken only with the advice and guidance of an architect or attorney. We recommend that the architect assemble the information and the attorney review it and from here on handle the communication with the contractor and the license board. Every case is different; therefore this opinion is very general, and just that—an opinion, not legal advice. These should be the next steps:

4. In the letter to the contractor, it would be stated that the contract should be considered terminated with a reference to the General Conditions, or a similar wording, namely, "because the contractor persistently refused or failed to supply enough and skilled personnel and/or proper material." We hope the A.I.A. General Conditions are part of your contract because the document represents the common position of the American Institute of Architects and of the Associated General Contractors of America. If the document is not part of your contract, care must be taken not to quote the General Conditions verbatim since they are copyrighted and must be bought and used as a complete document.

5. Solicit at least two competitive bids from other contractors to correct the faulty work and complete the project.

6. Let to one of these "secondary" contractors the contract, under a formal agreement.

7. Have the work inspected regularly by an expert. Notice of Completion, final inspections, punch lists and final payments will be made as under any regular contract.

8. After the secondary contractor is paid, the architect and attorney should also be reimbursed for extra work caused by the first contractor's failure to perform. They can be paid from funds originally due the first contractor, if there is money left. Any money still left over should be made available to the first contractor if he claims it. The A.I.A. General Conditions place an obligation on the owner to account for how the entire contract amount is spent, although the exercise normally will show a deficit, especially once an attorney must be involved to guide the termination. In this case, the owner might claim losses sustained through the action, or lack of action, of the first contractor.

Termination of a construction contract by either the owner or the contractor in most cases will not be clear-cut but will open the proverbial can of worms. Many questions not answered here will have to be resolved. For instance:

> What happens to the guarantee of work performed by the contractor who was terminated?
> Did the first contractor pay all labor and material bills, wage-related and other taxes?
> Was fund control switched?
> Did the lender accept the secondary contractor?
> What happens if it costs more to complete the project than what remains in the available construction loan?

DO:
- Hire the best contractor you can find and afford.
- Consider purchasing a Labor and Material Bond, and perhaps a Performance Bond, if you want to be especially safe. (See Chapter B-15.)
- Hire a good architect.
- Consult an attorney if (a) a mechanic's lien cannot be cleared up quickly, and (b) before terminating a construction contract.

CONSTRUCTION INDUSTRY ARBITRATION RULES

Effective January 1, 1981

**AMERICAN CONSULTING
ENGINEERS COUNCIL**

AMERICAN INSTITUTE OF ARCHITECTS

**AMERICAN SOCIETY OF
CIVIL ENGINEERS**

**AMERICAN SOCIETY OF
LANDSCAPE ARCHITECTS**

**AMERICAN SUBCONTRACTORS
ASSOCIATION**

ASSOCIATED GENERAL CONTRACTORS

**ASSOCIATED SPECIALTY
CONTRACTORS, INC.**

CONSTRUCTION SPECIFICATIONS INSTITUTE

**NATIONAL SOCIETY OF
PROFESSIONAL ENGINEERS**

**NATIONAL UTILITY CONTRACTORS
ASSOCIATION, INC.**

ADMINISTERED BY

AMERICAN ARBITRATION ASSOCIATION

140 West 51st Street   New York, N.Y. 10020

If arbitration becomes necessary, request a pamphlet from the American Arbitration Association.

You want to know more about a mechanic's lien. Let's talk about it next.

MECHANIC'S LIEN

The undersigned, _____ , claimant,

(Name of person or firm claiming mechanic's lien.
Contractors use name exactly as it appears on contractor's license.)

claims a mechanic's lien upon the following described real property:

City of _____ , County of _____ , California.

_____

(General description of property where the work or materials were furnished. A street address is sufficient, but if possible, use both
street address and legal description.)

_____

The sum of $ _____ together with interest thereon at the rate of _____

(Amount of claim due and unpaid.)   (See note on reverse)

percent per annum from _____ , 19 ___ , is due claimant (after deducting all just credits and offsets) for

(Date when balance became due.)

the following work and material furnished by claimant: _____ _____

(Insert general description of the work or materials furnished.)

_____

Claimant furnished the work and materials at the request of, or under contract, with, _____

_____

(Name of person or firm who ordered or contracted for the work or materials.)

The owners and reputed owners of the property are: _____

_____

(Insert name of owner of real property. This can be obtained from the County Assessor.)

**SEE REVERSE SIDE FOR
COMPLETE INSTRUCTIONS**

Firm Name ____ _____

(See instructions on back for proper signing.)

By : _____

(Signature of claimant or authorized agent.)

VERIFICATION

I, the undersigned, say: I am the _____ the claimant of the foregoing

("President of", "Manager of", "A partner of", "Owner of", etc.)

mechanic's lien; I have read said claim of mechanic's lien and know the contents thereof; the same is true of my own knowledge.

I declare under penalty of perjury that the foregoing is true and correct.

Executed on _____ , 19 ___ , at _____ , California.

(Date of signature.)   (City where signed.)

_____

(Personal signature of the individual who is swearing that the contents
of the claim of mechanics lien are true.)

Reproduced w/ permission from
WOLCOTTS FORM 1024—MECHANIC'S LIEN—Rev. 3-80   ©1980 WOLCOTTS, INC

Authors note: This form is carried by most good stationary stores throughout the West

C-13 MECHANIC'S LIEN

If an owner orders material, labor or services that are intended to improve his property and he disputes or does not pay the bill, the claimant has, under a state law, the right to file a mechanic's lien against the property to collect his money. A lien is a legal claim by one person on the property of another person for money owed. A mechanic's lien is similar to the way in which a deed of trust secures the payment of a loan. Within 90 days of filing the lien, the claimant has to file a foreclosure suit, otherwise the lien expires. Here is an example:

A brickmason made a written proposal to install a retaining wall and the owner signed it. After the wall was in place the owner refused to pay as agreed for three reasons:
- The back side of the wall was not waterproofed. A neighbor had told the owner there should be membrane waterproofing.
- The wall was not exactly where the owner, presumedly, had told the mason to put it.
- The mason had not left the place spic and span.

Without elaborating here whether the owner's reasons not to pay were valid, phony or doubtful, the upshot was the mason was mad and filed a mechanic's lien with the county recorder. Cost to the mason: $4.50.

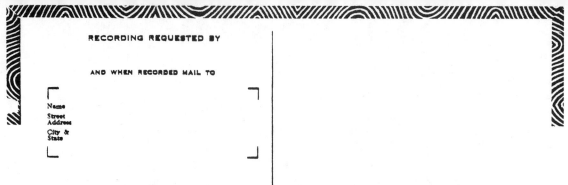

(SPACE ABOVE THIS LINE FOR RECORDER'S USE)

RELEASE OF MECHANIC'S LIEN

That that certain notice of lien executed by the undersigned against_____

and claiming a lien upon the following described real property situated in the
County of _____ , State of California, to-wit:

dated the _____ day of _____ , 19 ___ , and recorded in

the office of the County Recorder of _____ County on the

day of _____ , 19 ___ , in Book _____ of

page _____ as Instrument No._____ , is hereby released, the claim thereunder having been fully paid and satisfied.

WITNESS my hand this _____ day of _____ , 19

_____

_____

_____

STATE OF CALIFORNIA } SS.

COUNTY OF_____

On _____

before me, the undersigned, a Notary Public in and for said State, personally appeared_____

_____

_____

_____

known to me to be the person___ whose name_____

subscribed to the within Instrument and acknowledged that _____ executed the same.

WITNESS my hand and official seal.

_____
Notary Public in and for said State.

STATE OF CALIFORNIA } SS.

COUNTY OF_____

On_____

before me, the undersigned, a Notary Public in and for said State, personally appeared_____

known to me to be the_____President, and_____

_____ known to me to be the

_____ Secretary_____

the Corporation that executed the within Instrument, known to me to be the persons who executed the within Instrument, on behalf of the Corporation herein named, and acknowledged to me that such Corporation executed the within Instrument pursuant to its by-laws or a resolution of its board of directors.

_____
Notary Public in and for said State

Title Order No._____

Escrow or Loan No._____

This is not legal advice, just background information. **In order to file a mechanic's lien, at least two conditions must be met:**

1. In most states the claimant, except the general contractor, the architect or the contractors working directly for the owner, must have sent the owner, the general contractor and the lender a special notification form. In California it's called the "California Preliminary Notice." A sample is shown below. When receiving this notice the owner should not be offended; it is not a lien, just a notification that "if bills are not paid in full for labor, services, equipment or materials furnished or to be furnished, the improved property may be subject to mechanic's liens." Most professionals, contractors and vendors accompany this stiff form with a somewhat friendlier letter to deflate the impersonal language.

   Under the California Civil Code, a licensed contractor can be subject to disciplinary action if he fails to send the notice.

CERTIFICATE AND WAIVER OF LIENS

For the purpose of obtaining payment of the sum referred to on the face side of this Order and for the job referred to therein, the undersigned certifies that the labor and materials referred to therein have been actually furnished in the amount specified and were used on said job. All charges for labor and work done and materials purchased or furnished by the undersigned have been paid for by the undersigned in full. Upon payment of the sum specified on the face of this Order to the person to whom payment is directed and in satisfaction of any right of lien which the undersigned might have, the undersigned waives and releases any and all mechanic's lien rights upon the property referred to or to the improvements situated thereon and the right to file a stop notice on any construction funds by reason of work done, labor

performed or materials delivered prior to _____, 197____.

Any person who submits a false voucher to obtain Construction Loan Funds and doesn't use the funds for the purpose for which the claim was submitted is guilty of embezzlement. (Penal Code Sec. 484C)

Dated this _____ day of _____, 19____.

Signed _____   Signed _____
      MATERIAL SUPPLIER   SUB-CONTRACTOR

Typical WAIVER OF LIENS
This is the back side of the form ORDER ON FUND CONTROL,
shown on page 154 (Courtesy Western of San Diego Fund
Control Inc.)

Next page: California Preliminary Notice

189

CALIFORNIA PRELIMINARY NOTICE

California Society of
Professional Engineers

Consulting Engineers
Association of California

California Council of
Landscape Architects

C|C
L|A

California Council,
The American Institute
of Architects

Structural Engineers
Association of California

California Coun
of Civil Engine
and Land Surve

ORIGINAL CONTRACTOR or
Reputed Contractor, if any

————— FOLD HERE —————

OWNER   or   **PUBLIC AGENCY**
or Reputed Owner   **(on public work)**
(on private work)

————— FOLD HERE —————

CONSTRUCTION LENDER or
Reputed Construction Lender, if any

OWNER'S COPY

YOU ARE HEREBY NOTIFIED THAT . . .

_____

(name of person or firm furnishing labor, services, eq

_____

ment or material)

_____

(address of person or firm furnishing labor, services, eq

_____

ment or material)

has furnished or will furnish labor, services, equ
ment or materials of the following general descr
tion:

_____

(general description of the labor, services, equipment or

_____

terial furnished or to be furnished)

for the building, structure or other work of impro

ment located at:_____

(address or description of job site

_____

sufficient for identification)

The name of the person or firm who contracted
the purchase of such labor, services, equipment
material is:

_____

_____

(address of above person or firm)

If bills are not paid in full for labor, services, equi
ment or materials furnished or to be furnished, th
improved property may be subject to mechanic
liens.

Dated:_____

_____
(signature)   (title)

190

2. Definite time limits must be observed; these vary from state to state. In California:
    a. In general, a mechanic's lien must be recorded within 90 days after completion of work, if the owner has not yet filed a Notice of Completion.
    b. If the owner has filed a Notice of Completion (see C-14) or a Notice of Cessation of Work, a lien has to be recorded within: 30 days by subcontractors and vendors, 60 days by the general contractor.
    c. If the claimant is a contractor, supplier, etc., who has contracted not as a subcontractor with the general contractor, but **directly** with the owner, then he has the same right as a general contractor (who contracts directly with the owner). That means he has 60 days to record the lien.

Reading this far, any person of sound body and mind must be confused. Thusly, our government writes laws.

What happens after the lien is recorded and published? Not much, if the claimant does not file a foreclosure suit within 90 days.

The owner should try to settle the dispute with the claimant, pay the bill, or part of it, as agreed upon by both parties. When the case can be settled between the claimant and owner, but a lien was already recorded, it is best to remove the stigma of a lien which, through publication in local trade papers, becomes public knowledge. The owner should obtain a lien release from the claimant. In order to be valid a lien release must also be recorded.

While the mechanic's lien is on record, the claimant can charge interest on the unpaid balance. When no interest rate has been specified in the contract, it is 7 percent in most states.

DO:
- Avoid mechanic's liens! Pay architects, surveyors, consultants, contractors, tradesmen, etc., as your contracts provide.
- Obtain lien waivers from all contractors and suppliers.
- To avoid disputes, obtain written change orders for any extra work during construction that costs over $25.
- Have an arbitration clause in your contract; it can prevent lawsuits.
- State serious complaints against the contractor in writing, especially if you consider the complaint a reason for withholding payment.
- Take mechanic's liens seriously. Liens will be published and can play havoc with your credit and supply of material and manpower. Liens can delay your project at a much greater cost to you than the amount in dispute.
- Once a settlement is reached get a formal Mechanics Lien Release.
- Understand that lien laws vary from state to state, are complex and demand precise procedures, that affect the right to file a lien, the time limits and the priority of the lien over other liens and trust deeds or mortgages.
- If you cannot settle the dispute quickly, contact an attorney experienced in construction matters.

A much more appreciated document than
a lien is the Notice of Completion.
C-14 tells you why.

Almost there! . . .

C-14 NOTICE OF COMPLETION

You might have heard melodramatic tales such as the one about the homeowner who installed a swimming pool and had to pay for it almost twice. Although he had paid the pool company, some six weeks later he received bills, marked overdue, for ready-mixed concrete, ceramic tiles and unpaid workmen's compensation insurance. He also received a friendly but firm request from the labor commissioner for $465.73 for wages not paid to the cement finisher, Joe Smoe.

The owner threw up his hands—he had paid the pool people! But, you see, the pool people did **not** pay some of their suppliers or workmen. Under the law these parties have a right to demand payment directly from the owner who, in turn, has to go after the contractor. This is not a reflection on pool contractors in particular; most are competent and honest. But some swimming pools have been paid for twice, or nearly twice; the cases are on file.

It could happen to you, even if you don't build a pool. It can happen with any project; after you have paid the contractor, demands to pay off unpaid labor or material could fly through your mail slot. How? If any contractor with whom you had a direct contract did not pay any of his material suppliers, workmen or subcontractors and you find it impossible to make that contractor pay those bills, you are still liable. These parties can file a mechanic's lien or, as workmen, can ask the labor commissioner for assistance. After you have paid these overdue bills, you yourself have to go after the culprit, which in most cases is the general contractor.

If a mechanic's lien is filed and you don't pay, a lawsuit will have to follow the lien within 90 days; if a judgment is rendered against you, your building could go on the block and be auctioned off; from the proceeds the court officer will pay the debts and costs. Ach Du Lieber!

How do you avoid getting into such a situation? File a Notice of Completion! This is a one-page legal document to be filed with the County Clerk (County Recorder) at the time construction is completed. The owner, the lender, the architect or the contractor can file the notice. The lender may insist that he file it—fine; ask that you get a copy. Otherwise we prefer that the owner file it, or the architect on the owner's behalf. A sample of the

Notice of Completion form is on the next page.

There have been arguments as to when a building is "completed": When it is completed in substance (except for some paint touch-up, landscaping and adjusting a few door locks), or after the owner decided to change the entry door hardware for the third time and the Italian chandelier for the powder room arrived 12 weeks after the owner had moved in? Most construction contracts use the term "substantial completion." The General Conditions define the date of substantial completion as the date certified by the architect that construction is sufficiently complete in accordance with the contract documents so the owner can occupy or utilize the structure for the use for which it is intended. The Notice of Completion should be recorded within approximately 10 calendar days of the time of substantial completion.

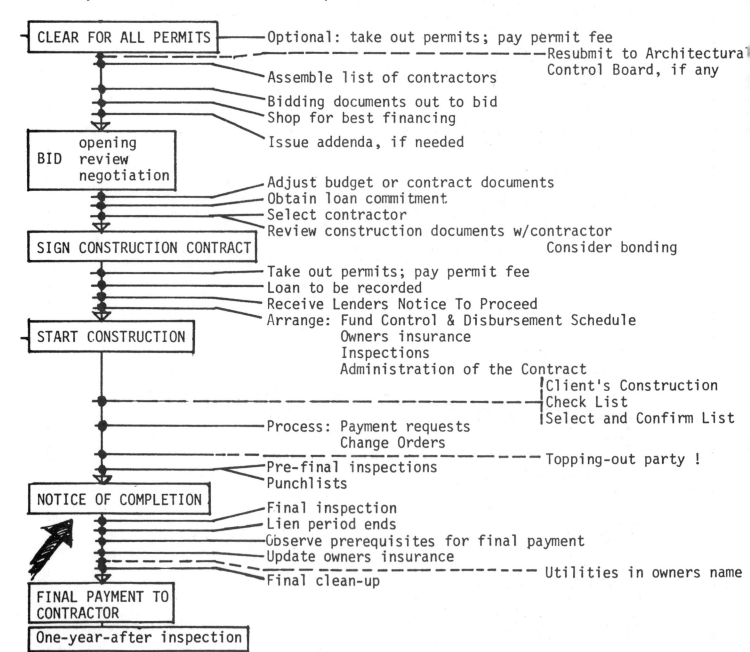

A NOTICE OF COMPLETION should be filed for all projects.

Order No.
Escrow No.
Loan No.

WHEN RECORDED MAIL TO:

SPACE ABOVE THIS LINE FOR RECORDER'S USE

NOTICE OF COMPLETION

NOTICE IS HEREBY GIVEN THAT:

1. The undersigned is OWNER of the interest or estate stated below in the property hereinafter described.
2. The FULL NAME of the undersigned is_____
3. The FULL ADDRESS of the undersigned is_____
   _____
4. The NATURE OF THE INTEREST or ESTATE of the undersigned is: In fee.
   _____
   (If other than fee, strike "in fee" and insert, for example, "purchaser under contract of purchase," or "lessee.")
5. The FULL NAMES and FULL ADDRESSES of ALL PERSONS, if any, WHO HOLD SUCH INTEREST or ESTATE with the undersigned as JOINT TENANTS or as TENANTS IN COMMON are:

   NAMES   ADDRESSES
   _____
   _____

6. The full names and full addresses of the predecessors in interest of the undersigned if the property was transferred subsequent to the commencement of the work of improvement herein referred to:

   NAMES   ADDRESSES
   _____
   _____

7. A work of improvement on the property hereinafter described was COMPLETED_____
8. The NAME OF THE ORIGINAL CONTRACTOR, if any, for such work of improvement is_____
   _____
   (If no contractor, insert "none.")
9. The street address of said property is_____
10. The property on which said work of improvement was completed is in the City of_____
    County of_____, State of California, and is described as follows:

Date:_____

Verification for INDIVIDUAL owner_____:

STATE OF CALIFORNIA
COUNTY OF _____ }ss.

* Signature of
owner named
in paragraph 2 _____
_____

The undersigned, being first duly sworn, states that ___he is the owner of the aforesaid interest or estate in the property described in the above notice; that ___he has read the same, knows and understands the contents thereof, and that the facts stated therein are true.

Signature of
owner named
in paragraph 2 _____

SUBSCRIBED AND SWORN TO before me

on_____

Signature_____
                    Notary Public in and for said state.

Verification for CORPORATE owner:
STATE OF CALIFORNIA
COUNTY OF _____ }ss.
_____

being duly sworn, says:
That he is the_____
of _____
_____, the corporation
that executed the foregoing notice as owners of the aforesaid interest or estate in the property therein described; that he makes this verification on behalf of said corporation; that he has read said notice and knows the contents thereof, and that the facts therein stated are true.

Signature of officer_____

SUBSCRIBED AND SWORN TO before me

on_____

Signature_____
                    Notary Public in and for said state.

Verification for PARTNERSHIP owner:
STATE OF CALIFORNIA
COUNTY OF _____ }ss.
_____

being duly sworn, says:
That he is one of the partners of_____
_____, the partnership that executed the foregoing notice as owner of the aforesaid interest or estate in the property therein described; that he makes this verification on behalf of said partnership; that he has read said notice and knows the contents thereof, and that the facts therein stated are true.

Signature of partner_____

SUBSCRIBED AND SWORN TO before me

on_____

Signature_____
                    Notary Public in and for said state.

(This area for official notarial seal) Form 1054(12/64)

What does the notice mean? The Notice of Completion will be published in local trade papers, and is a reminder to all who have supplied labor, material, equipment or services to your project that they have a certain number of days to let you or your lender know of any claims for payments they may have.

Thirty days or more after the Notice of Completion is recorded, the lender will release the final payment—either to the owner, to the general contractor directly or to fund control. If there is no lender, then the owner will make the final payment to the contractor. Both the lender and the owner have a stake in knowing which bills are outstanding at the time final payment is made; arrangements must be made to ensure that all bills will be paid, either through fund control before the final payment goes to the contractor or from that final payment itself.

Sometimes subcontractors and material suppliers don't want to wait until the general contractor receives the last payment so he can pay them. Remember the cabinetmaker: He installed kitchen cabinets and pullmans after drywall and exterior finishes, and if there is no "cabinet draw" (see C-4), he has to wait for payment until the lender releases the draw at Notice of Completion. By that time four to six weeks have passed. To add injury to insult, the cabinetmaker would get only a part of his money—because too many trades have to be paid from the "completion draw" the lender holds back the last 10 percent for 30 more days.

What is true for the cabinetmaker is true for other trades who work at that time—tile contractor, painter, carpetlayer—they not only have to wait for the lender's draw made at completion, but also another 30 days or more for a part of the payment.

There is a solution. The owner should arrange with the title insurance company for an indemnity agreement. If there is such an indemnity agreement the lender will release the last payment three or four days after the Notice of Completion is filed. By this device the owner (or fund control) is in a position to pay the subcontractors as soon as their portion of work is satisfactorily completed. Final payment to the **general contractor** shall be made only when the entire project is completed, down to the last doorknob, heating register, painters' touch-up and clean-up.

Yes, a Notice of Completion should be filed for all construction projects, large or small, new or remodeled.

DO:
- File a Notice of Completion or have the lender or architect do it.
  Where: County Recorder's office
  When: At "substantial completion"
  Cost: Approximately $3 to $4 for recording; the document will be sent to the lender or to you, as designated on the notice.
  Forms: Available from lending institutions, title companies, architects or office supply stores.
- Get yourself a copy of the filing. Cost: approximately $1.50.
- Be sure 30 calendar days have passed after the filing before making final payment to the general contractor unless there is an indemnity agreement with your title policy for the construction loan.

Permits, loans, contracts, bonds, liens
are important documents in the
process of building; they have legal
implications. There's one more:
Insurance! Sorry about that.

196

Painter falls off ladder . . .

C-15 INSUUURANCE!

- On a Saturday evening children started a bonfire in the wood framing of a house under construction. Damage: $1200.
- A garbage disposal, just installed by the plumber, was ripped out and never seen again.
- After the owner had selected the door hardware, medicine cabinets and bathroom accessories, the general contractor sent one of his men to pick up the material. When curving into the building supply parking lot, the driver skirted a Volkswagen and shaved off two fenders.
- One client had the painting work under a separate contract. When staining the exposed ceiling beams, the painter reached out too far, fell off the ladder and was laid up for three weeks. He is married, has four children and supports his mother.

Who pays for the framing?

Who pays for replacing the disposal? To repair the Volkswagen? To support the painter and pay for his medical bills? The answer in all these cases should be: The insurance. But whose insurance? And is there any?

Be patient. The subject of insurance can be an easy turn-off. To make it brief, almost too brief on this serious subject, the best advice can be summed up in three words:

HAVE SUFFICIENT INSURANCE!

When you build, have sufficient insurance, the right coverage and enough coverage. And the coverage must be in force as long as you build. This may sound obvious but you would be surprised to hear insurance people talk about how many contractors let their insurance lapse before the project is completed, or those who don't have enough coverage for the number of projects they work on. It doesn't hurt to remember the old saw: If nothing happens, you always have too much insurance; if something happens, you can never have too much.

Before we sell you on insurance, consider one more melodrama: We designed, built and sold a hillside house in San Diego Country Estates. Escrow was scheduled for

December 12, 1978. Our Course of Construction Insurance was good to December 20, 1978. We were safe. Then escrow was moved to January 2, 1979. Tax reasons, you know. Our broker called: "Get insurance to tide you over for those 12 days. If something happens you'll hate yourself forever!"

We took his advice and extended the insurance, for a fee of $20. Two days later, on December 21, a windstorm, the wildest in memory, hit the area. Winds up to 100 miles per hour flattened houses. A neighbor's roof and second story sailed into the hills. Fortunately, we suffered only slight damage in the path of disaster. Still, we relearned two lessons:

1. IT PAYS TO BUILD A STURDY HOUSE.
2. HAVE INSURANCE.

For most building projects the following insurance is recommended:

Type of Insurance	Risks Covered	*Min. Coverage Average Coverage	Paid By
Comprehensive general liability	contractors' liability; bodily injury; property damage	*$300,000 500,000	general contractor. If separate contract: subcontractor
Comprehensive automobile liability	bodily injury; property damage (by contractors' cars)	*$300,000 500,000	general contractor
Worker's Compensation	disability benefits; medical treatment	*approximately $150 per week; medical costs	general contractor or owner
Builders' risk (also called course of construction)	fire and extended coverage; vandalism; malicious mischief	replacement cost of building	general contractor or owner
Owner's contingent liability	bodily injury; property damage (for occurrences not the responsibility of the contractor)	*$300,000 500,000	owner
Performance bond	non-performance of contractor	job costs	owner
Labor and materials bond	non-payment of labor and materials	50% of job costs	owner
Title insurance with indemnity agreement	loss due to legal defects in title	*amount of loan replacement of cost	seller, developer or owner
Steam boiler and machinery, etc.	loss of equipment	amount of possible damage	owner

DO:
- Consult your insurance adviser for types, coverages and amounts of insurance you and the contractor should carry.
- Have a refund provision in the insurance contracts if you think you might not need the coverage for the full term of the policy.
- Request a certificate of insurance from each contractor **before** he starts working. Check the expiration date!

    The contractors insurance certificate must show that the owner is named as an additional assured.

- Arrange for worker's compensation insurance when hiring workers directly. Note: The worker's compensation part of the typical homeowner's policy normally does not cover construction work.
- Arrange for fire insurance with extended coverage as soon as framing begins. Have the contractor named as an additional assured to provide coverage for his interest in property on the jobsite.
- Include an indemnity agreement in your title policy.
- Consult with a reliable financial expert before buying mortgage insurance.

You need a . . .

C-16 PUNCHLIST

Near the end of the construction phase, the building inspectors make their final inspections, the Notice of Completion is filed and the contractor keeps talking about the final payment due. Easy now! Let the inspectors make their final appointed rounds. Having read this far you know who:

► The **Building Department inspectors** (structural, electrical, heating and plumbing) are mainly concerned with zoning matters and health, safety and energy-related requirements—for instance: Does the finish grade slope away from the building; is the 220-volt dryer outlet grounded; does the stair have uniform risers and treads; where is the house number; which windows need double-glazing?

► The **lender's inspector** normally signs off after the Building Department inspector, relying conveniently on their inspection records. He is not concerned about details as long as the building almost looks like the structure he sees on his set of drawings.

► **The inspector of the home owner's association,** if there is one, is satisfied that you didn't add a second story, since he saw your house in the framing stage; that the stucco color is light and the fascia russet brown, as you promised; and, "Please, always keep the garage door closed; this is an exclusive neighborhood."

► There could have been more final inspections, especially with commercial projects: from the fire marshal—exit lights, fire hydrants, extinguishers and sprinklers; from the **Engineering Division**—driveways, curbs, slopes; from the **Health Department**—commercial kitchen equipment, drains, finishes; and from the **Housing Department; F.H.A.; V.A.;** etc.

Why do we tell you this again? So you don't hang your hat on any of these inspections for a fast, complete peace of mind. You must do your own inspections to be sure the contractor performed everything called for in the contract documents (agreement, drawings, conditions, specifications, addenda, change orders).

Architect and owner should now start a series of inspections. We say "series," plural, because we can assure you that one final inspection will never do, not with a simple house remodeling, not with a 20-story high-rise hotel.

Make the first inspection when the project is "substantially completed." This happens when the contractor thinks so and the architect says so. Substantial completion was defined in Chapter C-14. If you have not had an architect or construction expert on your project so far, it would be prudent to hire one for final inspections.

Remember the Select and Confirm List and the Client's Construction Checklist? The information was organized in the sequence of construction and by trades because the **time sequence** in which work is performed is the natural format for those lists. Here then comes the third list from the architect to the owner. The list is compiled during the final inspection and is called the "PUNCHLIST." Punchlists are best organized in an **area sequence:** general items first, then room by room, and last the outside areas. Walk through the entire project and make notes of what is not done or completed but is part of the contract. That is the key phrase; it must be in your contract.

Tabulate shortcomings, first the general and typical ones for all areas, then list items room by room, starting at one end of the building. Below is the **first part** of a sample punchlist. As with all lists in this manual, the items listed are not typical for all projects; they occurred on one particular building. Typical is the arrangement of the punchlist. The length varies depending on the project, the quality of the contractor and the knowledge of the punchlister (if there is such a word). For a residential addition it might run three pages, for a new hotel, 20.

Punchlist for Sawyer Apartment

General: Manuals, guarantees, instruction leaflets, as-builts to owner.
Wash all windows, touch-up with bronze paint, remove labels.
Remove forming lumber from foundation walls, stoops, patios.
Paint tops and bottoms of all doors.
Some fluorescent fixtures hum, flicker, don't light up.
Touch-up with paint where carpetlayers damaged walls, etc.

Entry Hall: Install house numbers where marked.
Adjust door closer.
Three intercoms don't work.
Remove grout from floor tiles.
Paint-touch-up letterboxes, railing.

And so the list goes on.

As soon as the contractor says he has made all the corrections, have another inspection, punchlist in hand. What was not done properly stays on the list, the contractor has to come back, another inspection will follow. Don't give up.

Do not file the Notice of Completion yet. And do not make or call for the lender's next payment, which is normally due at substantial completion, UNTIL ALL MAJOR ITEMS AND MOST SMALL ONES ARE TAKEN CARE OF. If the powder room light fixture is still enroute from Paris or the Jacuzzi tub still sputters, these are small items. Most contractors welcome a realistic punchlist; they will instruct their men or subcontractors to comply with any reasonable request. The punchlist lets the contractor see the end of the tunnel. After all, the sooner the building can be completed the sooner the contractor can move to new—and maybe greener—pastures. Making the contractor realize that he has to perform his work before he receives the payments will bring a new spurt to the construction progress.

In cases where the general contractor will build the shell only, the punchlist should be issued as soon as he completes the shell. Later, punchlists should be written for the individual trades that work directly for the owner. The guarantee period for the shell starts at its substantial completion.

DO:

- Have a punchlist issued at the time of final inspections.
- Insist that most items be taken care of before the Notice of Completion is filed or the payment-due-at-substantial-completion is made.

APPLICATION AND CERTIFICATE FOR PAYMENT

AIA DOCUMENT G702

PAGE ONE OF   PAGES

TO (Owner):	PROJECT:	APPLICATION NO:	Distribution to: ☐ OWNER
		PERIOD FROM: TO:	☐ ARCHITECT ☐ CONTRACTOR ☐ ☐
ATTENTION:	CONTRACT FOR:	ARCHITECT'S PROJECT NO:	
		CONTRACT DATE:	

CONTRACTOR'S APPLICATION FOR PAYMENT

Application is made for Payment, as shown below, in connection with the Contract. **Continuation Sheet, AIA Document G703, is attached.**

The present status of the account for this Contract is as follows:

CHANGE ORDER SUMMARY

Change Orders approved in previous months by Owner	ADDITIONS	DEDUCTIONS
TOTAL		
Approved this Month		
Number	Date Approved		
TOTALS		
Net change by Change Orders		

ORIGINAL CONTRACT SUM . . . . $_____

Net change by Change Orders . . . . $_____

CONTRACT SUM TO DATE . . . . $_____

TOTAL COMPLETED & STORED TO DATE . . . . $_____
   (Column G on G703)

RETAINAGE _____% . . . . $_____
   or total in Column I on G703

TOTAL EARNED LESS RETAINAGE . . . . $_____

LESS PREVIOUS CERTIFICATES FOR PAYMENT . . . . $_____

CURRENT PAYMENT DUE . . . . $_____

The undersigned Contractor certifies that to the best of his knowledge, information and belief the Work covered by this Application for Payment has been completed in accordance with the Contract Documents, that all amounts have been paid by him for Work for which previous Certificates for Payment were issued and payments received from the Owner, and that current payment shown herein is now due.

CONTRACTOR:

By: _____ Date: _____

State of: County of:
Subscribed and sworn to before me this   day of   , 19
Notary Public:
My Commission expires:

ARCHITECT'S CERTIFICATE FOR PAYMENT

In accordance with the Contract Documents, based on on-site observations and the data comprising the above application, the Architect certifies to the Owner that the Work has progressed to the point indicated; that to the best of his knowledge, information and belief, the quality of the Work is in accordance with the Contract Documents; and that the Contractor is entitled to payment of the AMOUNT CERTIFIED.

AMOUNT CERTIFIED . . . . $_____
(Attach explanation if amount certified differs from the amount applied for.)
ARCHITECT:

By: _____ Date: _____

This Certificate is not negotiable. The AMOUNT CERTIFIED is payable only to the Contractor named herein. Issuance, payment and acceptance of payment are without prejudice to any rights of the Owner or Contractor under this Contract.

AIA DOCUMENT G702 • APPLICATION AND CERTIFICATE FOR PAYMENT • APRIL 1978 EDITION • AIA® • © 1978
THE AMERICAN INSTITUTE OF ARCHITECTS, 1735 NEW YORK AVENUE, N.W., WASHINGTON, D.C. 20006

G702 — 1978

This form is used by contractors to apply for progress payment for construction work.

C-17 THAT EXTRA MILE

In Chapter C-11, this advice was offered: in the construction cost breakdown treat painting work as an allowance and get bids on painting during construction. Then the owner and architect could decide whether to handle painting as a separate contract or to put it under the general contract by adding a margin, say eight percent, for the general contractor's coordination, supervision, risk and profit. The procedure sounds like a lot of extra effort on the part of the owner or architect, but is worth it for two reasons:

1. Painting contractors give better bids when they have a chance to walk with drawings and specifications in hand through the structure under construction and near completion. (The small amount of priming up to that point can be done by the general contractor.)

2. On many projects, the painting contractor will be squeezed. Since he is one of the last subcontractors to appear on the scene he is also one of the likeliest candidates for last minute budget cutting by the general contractor. But remember this: excellent painting work, staining and paperhanging can camouflage construction defects. Conversely, a mediocre painter, with substandard paints, can ruin an otherwise good building.

These are the practical reasons to handle painting in the bidding phase as an allowance item and sometimes as a separate contract in the construction phase.

But there is also a symbolic aspect to this advice: a plea that the client, for his benefit, allow for some special effort that will save the building from being dull or mediocre, the proverbial pea in the pot of peas. A plea that in the planning stage more than routine consideration will be given to the finished appearance of a building.

Go that extra mile, we encourage clients. Plan, construct and finish a building in a strong, definite way. Use good materials—not goldplated, just good. Use special features that are functional as well as pleasing. For instance: change in floor levels and ceiling heights, interior courts, see-throughs, moving water, French doors, skylights, window seats. At the same time use restraint in the number of colors, textures, materials. Install light fixtures, door hardware, railings, furniture, fences of simple design but substantial quality.

Many new church buildings are unfortunate examples of architect's frustrations: the building committee could have gone that extra mile. After finally deciding to build the new sanctuary—or enlarge the fellowship hall—after all those countless meetings, after spending tens of thousands of churchgoers' dollars, it will be easy to detect long faces at the opening ceremonies.

Where is the warmth, the strength—the feeling of being in a special place? Everything seems flat, hollow, thin, plastic, colorless, assembled without love. Where is the play of light and shadow? Did anyone pay attention to acoustics? Street noise filters through thin doors and aluminum windows. Not enough effort was made in the planning phase. Now this disappointing church structure, this turkey will stand there for the next 60, 80 or 100 years, not a place to worship but a place to cry.

Warmth, flair, quality, intelligent planning pay. They provide easy maintenance, cut repair costs, give lasting satisfaction. Simplicity and imagination, strength and quality make sense and make a good return on the investment. This is true for all buildings—log cabins, residences, apartments, office towers, churches, airport terminals.

DO:

- Build quality into your project. It pays.

From a Letter to the Editor
of a national magazine:

"It is obvious to many of us that there
never will be a demand for good planning and design
by the general public unless these subjects are
introduced in the public schools and continued through
our universities, especially in the colleges of education
and of arts and sciences."

> P.M. Torrace
> Professor of Architecture Emeritus
> University of Florida.

C-18 BEFORE MAKING THE FINAL PAYMENT

Normally, the final payment to the (general) contractor is made

> either from the lender to fund control or
> from the lender directly to the contractor or
> from the owner to fund control or
> from the owner directly to the contractor.

Regardless who makes the final payment to whom, it is crucial to the owner's interest that certain requirements be met by the contractor.

Human nature being what it is, it's more difficult to have these requirements fulfilled once the final payment has been disbursed. These requirements are included in the "General Conditions of the Contract for Construction," A.I.A. Document 201, and are a part of the contract between the owner and the contractor as long as the General Conditions are included in the contract.

If the General Conditions are not made part of the contract, it is best to enumerate the prerequisites for final payment either in the contractor's proposal form (which becomes a contract when signed by owner and contractor) or in the contract form itself. You have been cautioned against using just any form the contractor might submit to you; some of these contract forms not only have sizable loopholes, but also are slanted in favor of the contractor. They don't, for instance, nail down the final payment requirements.

Here are the prerequisites for final payment:

1. Notice of Completion must be filed.

2. All work must be completed; not only substantially, but entirely and to the satisfaction of the architect.

3. Owner must receive:
   a. Permit set, records of inspection, as-built drawings (if part of the contract), certificate of occupancy (not required for residences)
   b. Manuals of instruction, instruction of owner's personnel with systems operation and maintenance
   c. Manufacturers' and contractors' guarantees

d. Updated list of subcontractors and suppliers, including addresses and telephone numbers

e. Keys, keying schedule, change of locks, tools, spare parts and replacement materials (if part of the contract)

4. All governing agencies must conduct final inspections and sign off the inspection records. These include, of course, the building inspection department and sometimes the engineering department (grading work on public property), the department of housing (apartments), home owners associations (CC&R's), and the fire marshal (on larger projects).

   For all buildings except single family residences, a Certificate of Occupancy must be issued. Normally the inspection department will send it directly to the owner, approximately two weeks after final inspection and sign-off. If the contractor built the shell only, a Certificate of Completion will be issued for the shell and once the building is completed the Certificate of Occupancy will be issued.

5. Excess funds are returned to the owner either by check or by deducting the amount from the final payment due. The owner might also be entitled to a refund from the engineering department for unused inspection fees and/or bonds on deposit. This money will not be returned by the contractor but by the respective city or county department directly to the owner since he made the deposit to these agencies.

6. To avoid mechanic's liens, arrangements must be made by which all subcontractors, material suppliers, workers and taxes are paid by the general contractor before or simultaneously with the final payment. This can be done in two ways: either by requesting a list from the general contractor of all outstanding bills (this list is then sent to fund control), or by the general contractor submitting an affidavit stating that all parties have been paid and lien waivers have been received. The waivers of lien shall be given to the owner.

7. A minimum of 30 calendar days must elapse between filing the Notice of Completion and final payment, unless the owner has an indemnity agreement with the title company. The lender probably wants to see the indemnity agreement. (C-14)

   Contractors expect to be paid in full about 30 days after the Notice of Completion is filed. If they are not paid without sufficient cause, contractors can charge interest at 7 percent per annum, or whatever their contract states, and file a mechanic's lien on the property. If some work is not completed by the time the final payment is due, the owner should retain a portion of the final payment. The retainage must be sufficient that another contractor could complete the work if necessary.

DO:
• Require the contractor to fulfill certain contract obligations before making the final payment.

C-19 11 MONTHS AND 29 DAYS LATER

Like most manufactured products, a new building should be guaranteed against faulty workmanship and material for one year. With new houses in subdivisions or planned residential developments the guarantee is taken for granted by the home buyer. But no guarantee is automatic; the buyer should look for the one-year guarantee in the purchase agreement. There also should be a clause to the effect that any defects brought to the attention of the seller, but not taken care of satisfactorily in the first year or immediately thereafter stay under the guarantee until repaired, whether the claimant is the original buyer or not. Requests for repairs under the guarantee should be in writing.

What about guarantees for custom residences, commercial buildings or remodeling projects? Same as above: should be guaranteed for one year; guarantees should be spelled out in the contract between the owner and the contractor. The place for such a guarantee clause is, in order of preference:

1. In the agreement between owner and contractor.
2. In the general conditions or special conditions. Both are part of the specifications.
3. On the standard notes sheet of the drawings.

A minimum note that would serve the purpose is as follows:

> General contractor and subcontractors shall individually guarantee
> for one year all materials and workmanship except as otherwise
> agreed to in writing.

Architects often stipulate guarantee periods different than the typical one-year guarantee.

Roofing should carry a two-year guarantee.

Structural defects like the sagging of foundations or bulging of retaining walls are covered by the statute of limitations, which vary from state to state. On the other hand some manufactured equipment, exhaust fans for instance, carry only a 90-day guarantee by the manufacturer and that is what the general contractor probably will give the owner.

To ensure the benefit of the guarantee the owner should schedule a guarantee-period inspection approximately 11 months after the Notice of Completion is filed. The procedure is similar to the final inspection at the close of the construction phase:

• A competent person inspects the building with the owner and the general contractor.

A list of required corrections (punchlist) is written. The architect is the best qualified for this—he knows the contract and the building. He is also best equipped to function as an arbiter between owner and contractor if small disputes evolve.

- The contractor makes the corrections or repairs; let's hope he is still in business a year later. Yes, there is another valid reason for hiring a reputable contractor in the first place.
- After repairs are completed: another inspection, and maybe a third. Never should the owner give up on getting things straightened out or replaced; and he should leave it to the architect to decide when everything is in order. The architect, of course, will be paid for this service, either a flat fee or by the hour. The exercise is worth the expenditure.

Two notes of caution:

One: The owner can only expect to receive what is in the contract. If unseasoned lumber was permitted for a ceiling with exposed beams, it will check and split and crackle when the temperature changes. Where cheap exhaust fans were specified, they will stay noisy forever. If joists under bathtubs are longspan and wide-spaced, the tile joints at the tub rim will open up, and stay open, etc. *Don't try to use the guarantee inspection to upgrade your building posthumously!*

Two: A reputable general contractor will come back and make good on repairs. But in case the Wunderbar Construction Co., Inc. is not in business anymore and its former president sells real estate now or pots and pans in another state, the owner might be up the proverbial creek. He can try to contact the subcontractors directly and the good ones may respond. This is the primary reason for getting a list of subcontractors before making the final payment to the contractor. Try to contact any subcontractor directly when repair work is needed and the general contractor is gone with the wind. Legally, the owner never had a contract with any of the subcontractors; their contract was with the general contractor, and every sub who is in business longer than three months knows that. He does not have to respond except if he gave the owner directly a written guarantee.

The bottom line is: does the owner have any leverage to force an unwilling contractor back on the job in the first year after construction? He has, depending on how much of a gadfly the owner chooses to be. Our advice: Don't give up easily. In case of difficulty:

► Have the architect or your attorney write a registered letter to the contractor.
► File a Consumer's Complaint. Get the special form shown in Chapter C-12 from your local office of the Contractors Registration Board. Send a copy of the complaint with a brief note to the local Better Business Bureau.
► If the contractor was or is a member of the local builders association go there personally and ask for advice.

DO:
- Conduct a year-after inspection and request from the contractor repairs or adjustment as necessary. If you hired a reputable contractor to begin with, we almost can assure you he will be back.
- Consider having subcontractors give the owner written guarantees directly. That creates a contractural relationship between owner and subcontractor at least at the end of the project. Such written guarantees must be specified in the specifications.

Check your project against every
recommendation of the Cost-Shaving
Checklist that follows.

Refer to	DO	Importance to reduce construction costs, hazards or liabilities	Pertains to your project	Done
C-1	Pick the best time to build.	■ ■ ■		
C-1	Have construction contract in writing.	■ ■		
A-17, C-2	Read all contract documents. Twice! Ditch all outdated drawings.	■ ■		
C-2	Check expiration dates of permits.	■		
C-2	Press for shortest possible construction time to save interest on the construction loan.	■ ■ ■ ■		
C-15	Know your insurance obligations.	■ ■		
	Verify contractors' insurance.	■ ■		
	Have refunding provisions with all insurance policies.	■		
C-3	Decide who pays for utilities and job toilet.	■		
C-4	If not already a condition of your lender, consider fund control.	■ ■		
C-6	Purchase anything you buy directly at "dealers' cost", through the contractor.	■ ■ ■		
C-6	Make use of your architect's resale license.	■ ■		
C-7	Have the project continuously checked by a competent person other than the contractor.	■ ■ ■		
C-9	Maintain good relations with inspectors.	■ ■		
C-10	During construction: take names of good workmen you might want to hire later for additional work; obtain consent of contractor. Arrange insurance for these workers.	■		
	Request from architect:			
C-5	Select and Confirm List	■ ■		
C-8	Client's Construction Checklist	■ ■		
C-16	Punchlist	■ ■		
C-13	Avoid mechanic's liens.	■		
C-14	Have Notice of Completion filed.	■ ■ ■		
C-18	Obtain from contractor: as-builts, manuals, instructions, guarantees, refunds from allowances and/or funds deposited with government agencies, keys, lien waivers.	■		
C-19	Specify and conduct a "one-year-after" inspection.	■ ■		

Refer to	DON'T			
C-1	Start construction before certain conditions are met: Building loan approved, written contract with contractor, certificates of insurance received, also: building permits and clearance from architectural control board, if any.	■ ■		
C-4	Pay any advance to any contractor for any reason.	■ ■ ■		
C-12	Hesitate to terminate an incompetent architect, consultant or contractor.	■ ■		
C-18	Make final payment to contractor before he has fulfilled certain requirements: completed entire work, filed Notice of Completion, refunded balances of allowances, arranged final inspection, turned over manuals, guarantees, etc.	■ ■ ■		

P.S. Yes, there are exceptions, but darn few!

FAMOUS LAST WORDS

What does it take to make your building project a success? It takes three steps.

First: Decide to do it.

Second: Gather information, then understand the information, the risks, the details, the alternatives.

Third: Act! With dedication and prudence, persistence and grace.

            DO IT — LIKE NOTHING ELSE MATTERS.

Good Luck and Farewell.

PART D: APPENDIX

D-1 WORDS YOU ARE GOING TO USE AND WHAT THEY MEAN

The definitions are general, nontechnical where possible, and brief. They do not encompass all meanings that a term may acquire in a legal context.

Examples are given in parentheses.

Addendum:

Written modifications or interpretation of drawings and/or specifications **after** they have been delivered to bidders but **before** bids have been received; to be issued by architect or owner to all bidders. An addendum becomes part of the construction contract.

Administration of the Construction Contract:

Final phase of the architect's basic services; sometimes incorrectly called "supervision." (The contractor supervises, the architect administrates the contract.)

A.I.A.:

American Institute of Architects.

A.I.A. forms:

Very useful contract and construction management forms copyrighted and issued by the A.I.A. May be used by anyone who buys the forms. Available from A.I.A. chapter offices and some blueprint companies.

Allowance:

Sum allocated in Contract Documents, for items or services still not specified at time of bidding (allowance for ceramic tiles).

Alternate:

Substitute choice.

Alternate bid: The cost of a specified selection that might take the place of another, not necessarily equal in appearance, quality or price.

("Give alternate bid for plaster in lieu of drywall.")

Approved equal

A substitute considered equal but unknown or not yet selected when specifications are written; to be approved by architect, engineer or owner.

("Joist hangers to be Simpson or approved equal.")

Arbitration:
Settlement of a construction dispute by a group of experts, the Arbitration Board, chosen to hear both sides and come to a decision.

As-Builts:
Drawings or prints marked up to reflect important changes made during construction. Furnished to architect and owner by the contractor **if** specified as a condition of the owner-contractor agreement.

Base Bid:
Bid on the contract **not** including alternates.

Bid:
Single Bid or negotiated contract: Arrived at during the negotiation between owner and contractor.
Competitive bidding: Several contractors, virtually in competition with each other, submit bids.

Bidding and Negotiating Phase:
In the contract between owner and architect the fourth phase of architects services. The phases are:

	In this book
Schematic design	
Design Development	Part A
Construction Documents	
Bidding and Negotiating	Part B
Construction Administration	Part C

Bidding Documents:
Consists of:
Invitation to bid (seldom used, except on public work)
Instructions to Bidders
Proposal Form
Drawings
Specifications,
Conditions
Addenda
Sample of Owner-Contractor Agreement
Sometimes included: a form to list proposed subcontractors and vendors

Blueprint:
Reproduction of an original drawing or of a reproducible sepia; has either white lines on blue paper or blue, brown or black lines on white paper.

Board of Appeals:
Group of experts, appointed by local elected officials, to hear and examine:
► complaints about interpretation of codes
► construction methods considered by an engineer or architect as equivalent to those prescribed by codes.

Bond:
Written obligation to pay a specified amount.
Labor & Materials Payment Bond: guarantees that all labor and materials will be paid.
Performance Bond: guarantees that work will be performed per Contract

Documents. In states where permitted: often combined with Labor & Materials Payment Bond.

Bonus & Penalty Clause:
Stipulation in the Contract Documents to provide a bonus payment in case the work is completed prior to the agreed completion date and a penalty charge when completed after that date.

Building Department Corrections:
A list of required corrections and/or information needed or to be clarified and to be supplied by the owner, his agent or architect before a permit can be issued.

Building Designer:
In California a person licensed by the state to design certain buildings, produce the Construction Documents and administrate the contract in a similar manner as an architect.

Building Official:
Person or his authorized deputy charged with the administration and enforcement of a building code; includes building inspectors of the city, county, or state.

Building Permit:
Written permission to construct or alter a structure, in most cases supported by approved drawings and specifications. Has time limits.

Building Zone:
An area within boundaries as defined by planning and zoning regulations. Building zones are established in order to regulate, restrict and segregate the location of industries, businesses, trades, apartments, single family dwellings and other specified uses. Regulations are available from the zoning departments of municipalities.

Certificate of Occupancy:
To be issued by the building official for most types of buildings, except single family residences, before the building can be used or occupied. When the existing occupancy classification changes a new certificate is required.

Change Order:
Instruction to contractor, **after** contract is signed, to change work, construction time or costs. It should be in writing if the change costs more than $25 and be signed by owner, contractor and architect.

Conditional Use Permit:
Planning commissions are authorized by most municipal codes to issue Conditional Use Permits for special land uses that are not included in the normal range of permitted uses in any zone. A pamphlet describing the procedures can be obtained from city or county administrations.

C.C.&R.'s:
Covenants, Conditions and Restrictions. Binding guidelines for the design, erection and maintenance of structures. Normally issued by the developer or a home owners association. Has the legal implications of a private contract.

Construction Cost Breakdown:
List of all building and related costs, broken down by trades, items, overhead, expenses and allowances. Obtain the form from your lender since every lending institution uses its own form.

Contingency Allowance:

Sum allocated for miscellaneous, unspecified or unforeseen work, items or services (blasting of rock to accommodate foundation trenches). Monies from this allowance can be spent only with owner's consent; unused portions shall be returned to the owner.

Contract Documents:

Consists of:

Agreement between owner and contractor

Drawings

General and Special Conditions

Specifications

Addenda

Change Orders, X-drawings

Any other document specifically designated in the Agreement as being a Contract Document

Contractor:

Licensed organization or person, committed by contract to perform certain construction work, as defined by Contract Documents.

General Contractor: Has single responsibility for all workmen and subcontractors working under him.

Subcontractor: Has contract with general contractor to perform certain part of the work (carpentry, heating, etc.); has no contractual relationship with owner.

Cost of the Work Plus Fee Contract:

A construction contract by which the owner pays the general contractor the cost of all subcontracts and jobsite overhead plus a fixed fee for managing the project. Subcontractors may still be selected by competitive bidding.

Course of Construction Insurance:

Fire Insurance with extended coverage and Liability Insurance taken out by the owner during the construction period; can be converted to regular Fire Insurance or Home Owner's Insurance when the building is sold, rented or inhabited by the owner.

Courtesy Bid:

Contractor's proposal based on a superficial and high cost estimate. A courtesy bid is not made to get the contract but merely to comply with the owner's request for a bid. A waste of time for both parties. Should be discouraged.

Coverage:

The portion of the area of a lot, expressed as a percentage, occupied by all buildings and structures that are roofed or otherwise covered and which extend more than 3'0" above grade level. Maximum allowable coverage is determined by zoning regulations and vary from building zone to building zone. Sometimes hard-surfaced areas such as parking or concrete walks are counted as partial coverage.

Draw:

Partial payment made by the lender toward the construction loan to owner, contractor or Fund Control.

"Dutch" Interest:

A very high interest configuration on the construction loan whereby the interest is computed on the total loan amount from the minute the loan is *recorded.* Recorded!—Not when money is drawn! This is considered an immoral financial condition imposed by a lender. Stay away from loans with "dutch" interest.

Encroachment Permit:
Issued to allow construction on neighboring public property. (A marquee that is attached to a building and projects over a public street.)

Escrow:
A third party's custody of papers, money and instructions until the wishes of two or more parties are carried out. "Escrow holder" can be a financial institution, title company, escrow company or attorney.

Fast Tracking (Construction Lingo):
To plan and erect a building in the shortest possible time. The design, preparation of construction documents, financing, bidding and construction are done in overlapping phases.

Final Inspection:
Review of the project before final payment, best conducted jointly by architect, owner and contractor. Normally followed by a final Final Inspection, after the items on punchlists are completed.

Final Payment:
Final payment to contractor will be made by the owner after architect issues Final Certificate for Payment or similar instructions. Final payment is not necessarily the unpaid balance of the contract sum, since the final payment might be adjusted by any of the following: change orders, return of unused portions of allowances, deductions for uncorrected work, penalties and bonuses, deductions for liquidated damages or for reinspection payments.

Finishing Trades:
The trades that work after the "shell" of the building is completed: drywall, interior plaster, finish carpentry, ceramic tiles, suspended ceiling, flooring, painting, carpet laying, etc. Although plumbers, electricians and heating subcontractors have finishing phases, they are not considered finishing trades.

Fire Zone:
Designation by the fire marshal or the building inspection department, given to every area within a city. Normally:
    Fire Zone 1 = Downtown
    Fire Zone 2 = Commercial Areas (C Zones)
    Fire Zone 3 = Residential
Fire-protective design standards for buildings differ with each fire zone. Verify fire zone before designing any structure.

Floor Area Ratio:
A planning department's device to control density expressed as a percentage. The allowable Floor Area Ratio varies from one building zone to another. (Allowable F.A.R. of 0.6 means the area (in square feet) of **all floors** of the building can be 60 percent of the lot area.)

Fund Control:
Disbursements of construction funds from the lender to the contractors or owner-builder through a third party, a fund control service, which is regulated by state laws, similar to an escrow company.

General Conditions:

"General Conditions of the Contract for Construction" Copyrighted A.I.A. document A-201 defines obligations, rights and relationships of the parties involved in a construction contract.

Hard Costs:

The construction costs to a contractor up to the point where his own overhead, taxes, contingencies, risk and profit are added.

Indemnity Agreement:

Contract by which one company agrees to reimburse another company for a loss. (Title insurance company will reimburse lender for any losses caused by lender's making final payment before end of lien period.)

Labor & Materials Bond:

See Bonds.

Lumpsum Contract:

See stipulated sum contract.

Mechanic's Lien:

Encumbers real property for the amount of unpaid labor, materials or services. Until these claims are settled, a clear title to the property cannot be given. Lien laws are complex and vary from state to state.

Notice of Completion:

Legal document filed with the County Recorder by the owner or by the architect or lender on the owner's behalf. The notice is filed at substantial completion of the project within 10 days after work is substantially completed. Notice starts the period during which mechanic's liens still can be filed.

Originals:

Original drawings, also called "tracings," as opposed to prints.

Package Builder:

A firm that furnishes design, contract documents and construction services under one contract. Sometimes additional activities such as lot selection, financing and building maintenance are also part of the design and building package.

Performance Bond:

See Bonds.

Pick-up Work:

The 101 tasks that are part of the general contractor's obligations but, following a sloppy tradition, are never assigned to any particular subcontractor. For instance: installing finish hardware and the pushbutton for the door chime, adjusting gate hinges, removing debris, vacuuming heating ducts, touching-up walls with paint after the carpet has been installed, etc., etc. The ideal contractor's pick-up man is a combination carpenter-housemaid-handyman-truck driver-janitor. During construction he also shows the half-finished building to all uninvited drop-ins.

Point (Lender's Lingo):

Sometimes called "discount points." Means one percent of the amount of a loan. (One point of a $60,000 loan is $600. Points are charged by lenders to raise the yield on loans.)

Progress Payment:
  Partial payment made by owner, lender or fund control to contractor.

Punchlist:
  List of required corrections compiled after semifinal and final inspections. Normally prepared by the architect and given to the contractor. Items on punchlists must be corrected or complied with before final payment to contractor is made.

Quantity take-off:
  List of all materials and equipment needed for a construction project.

Reinspection Fee:
  Charged by some building inspection departments for having to make a third or fourth inspection of construction work that should have been corrected at the time of the first reinspection—which was the second inspection of the same item!

Release of Mechanic's Lien:
  Legal document to be filed with County Recorder after Mechanic's Lien was filed but then the dispute settled. To be executed by the party who filed the lien.

Rendering:
  Artist's concept of a proposed or finished project, in black and white or color, sometimes called "perspective."

Retainer:
  A sort of earnest money charged by architects when the owner-architect agreement is signed. Retainer will be refunded to owner by a deduction from the last payment to architect.

Schedules of Values:
  See construction cost breakdown.

Scope of Work:
  Introductory statement in specifications, to enumerate what the specifications cover: labor, material, work included, related work not included, installation, workmanship, cleanup and guarantee requirements.

Separate Contract:
  Contract directly between owner and a specialty contractor when the owner also has a contract with a general contractor. No contractual relationship exists between specialty and general contractor. (Heating contractor deals directly with owner and is not one of the general contractor's subcontractors.)

Shell of a Building:
  A vague term that includes the work of the trades preceding the finishing trades, namely: grading, foundations, rough plumbing, electrical, heating, framing, masonry, roofing, exterior plaster or siding, exterior doors and windows.

Shop Drawings:
  Detail drawings, prepared by subcontractors or equipment suppliers, to be checked by general contractor and approved by architect. ("Layout for heating, cooling and venting" by heating contractor.)

Special Conditions:
  Section of the specifications pertaining to conditions and requirements unique to a particular project and not typical enough to be a part of the general conditions or standard notes. (For a store remodeling the special conditions might spell out

how the store operation is being maintained during construction.)

Specifications:
Part of the contract documents containing detailed written description of materials, installation, workmanship, standards and guarantess; organized in 16 sections. Normally typed on 8½ x 11-inch sheets and bound into a folder.

Square Foot Cost:
Assumed average construction cost per square foot of a proposed structure, derived from actual costs of recently completed similar construction. After the building is completed: the actual average cost per square foot of that structure.

Standard Notes and Details:
A drawing with general notes, typical Building Department requirements and standard details.

Statute of Limitations:
A state law limiting the time in which court action can be brought to recover losses from faulty building construction. Varies from state to state.

Subcontract:
A contract between general contractor and subcontractor; does not create a contractual relationship between owner and subcontractor.

Sub-subcontractor:
Contractor who has a contract with a subcontractor but has no contractual relationship with general contractor or owner. (Concrete pumping contractor, working for concrete subcontractor.)

Substantial Completion:
When a building is completed to a degree that the owner can take occupancy, normally after the first final inspection and when only minor corrections still have to be made. "Notice of Completion" should be filed at this time.

Subsurface Conditions:
Allowable soil pressure, type of soil, presence of fill as revealed by a soil investigation and documented by a soil report.

Superintendent:
The contractor's representative on a construction project. Oversees jobsite operations, coordinates trades, orders materials, interprets drawings and specifications, keeps contact with architect and/or owner, is responsible for safety measures, arranges for as-built drawings.

Supervision:
By contractor: Direction of the work at the jobsite by general contractor's personnel, normally the construction superintendent or a foreman.
By architect: A misnomer—he does not supervise, the contractor does. The architect "administrates the construction contract."

Survey:
A map drawn by a licensed surveyor showing the results of measuring the land with its elevations, boundaries, trees, improvements and its relationship to surrounding properties.

Time and Material:
A very loose contract arrangement where a contractor performs work and bills the

owner accordingly, often with no competitive subcontractors' bids and no guaranteed maximum cost. Not to be confused with "cost of the work plus fee" contract.

Title Insurance:

Protects lender or owner against loss of their interest in property due to legal defects in title.

UBC

Uniform Building Code. Provides mandatory minimum standards for the design, construction, use, occupancy and maintenance of buildings and related structures.

Unit Price:

Cost quotation of a unit of work as described in drawings or specifications, and stated in bids or the construction contract. ("Additional excavation and relocating the excavated material at the site, as directed in the field, will be $7.20 per cubic yard.")

Variance (Zone Variance):

A formal permission granted by the zoning administrator after a hearing to modify existing zoning provisions for the benefit of the applicant.

Working Drawings:

Plans, sections, elevations and details, drawn to scale and dimensioned. Also schedules and notes, all produced in the construction documents phase, after preliminary drawings are approved by owner. Working drawings will be used for cost estimates, approval by government agencies, bidding, final loan arrangements and construction.

X-Drawing:

A drawing issued by the architect after the signing of the owner-contractor agreement. Normally issued as part of a change order; both change order and X-drawing then become contract documents.

D-2 STANDARD ABBREVIATIONS

@	AT		D.G.	DECOMPOSED GRANITE
A.B.	ANCHOR BOLT		DIM.	DIMENSION
ACOUS.	ACOUSTIC		DN.	DOWN
ADJ.	ADJUSTABLE		D.S.	DOWN SPOUT
ALUM.	ALUMINUM		DTL.	DETAIL
ALT.	ALTERNATE		D.W.	DISHWASHER
APP.	APPROVED		DWG.	DRAWING
ASPH.	ASPHALT		EA	EACH
BLDG.	BUILDING		E.P.	ELECTRIC PANEL
BLK'G.	BLOCKING		ELEC.	ELECTRIC
BM.	BEAM		ELEV.	ELEVATION
BRK.	BRICK		EQ.	EQUAL
CA.	APPROXIMATELY		EXIST.	EXISTING
CAB.	CABINET		EXP.	EXPANSION
CEM.	CEMENT		EXP. JT.	EXPANSION JOINT
CER.T.	CERAMIC TILE		EXT.	EXTERIOR
C.J.	CEILING JOIST		F.A.U.	FORCED AIR UNIT
CL.	CLOSET		F.D.	FLOOR DRAIN
C_L	CENTER LINE		F.G.	FIXED GLASS; FUEL GAS
CLG.	CEILING		FIN.	FINISHED
C.M.U.	CONCRETE MASONRY UNIT		F.J.	FLOOR JOIST
COL.	COLUMN		FL.	FLOOR
CONC.	CONCRETE		F.O.C.	FACE OF CONCRETE
CONST.	CONSTRUCTION		F.O.M.	FACE OF MASONRY
CONT.	CONTINUOUS		F.O.S.	FACE OF STUD
CONTR.	CONTRACTOR		FTG.	FOOTING
C.T.	CURTAIN TRACK		GA.	GAUGE
C.W.	COLD WATER		GALV.	GALVANIZED IRON
ϕ	DIAMETER		GEN.	GENERAL
DBL.GL.	DOUBLE GLAZED		G.D.	GARBAGE DISPOSAL
D.F.	DOUGLAS FIR		G.I.	GALVANIZED IRON

GL.	GLASS	RDWD.	REDWOOD
GR.	GRADE	REC.	RECESSED
GYP. BD.	GYPSUM BOARD	REINF.	REINFORCED
H.C.	HOLLOW CORE	REQ'D.	REQUIRED
HDR.	HEADER	RES.T.	RESILIENT TILE
HDWE.	HARDWARE	R.J.	ROOF JOIST
HGT.	HEIGHT	RM.	ROOM
HORIZ.	HORIZONTAL	R.O.	ROUGH OPENING
HTG.	HEATING	R.S.	RESAWN
H.W.	HOT WATER	R.V.	ROOF VENT
I.D.	INSIDE DIAMETER	S.C.	SOLID CORE
INSUL.	INSULATION	S.C.R.	SHOWER CURTAIN ROD
JST.	JOIST	SECT.	SECTION
JT.	JOINT	SHT.	SHEET
LAM.	LAMINATED	SHO.HD.	SHOWER HEAD
LAV.	LAVATORY	SL.GL.	SLIDING GLASS
L.P.	LOW POINT	SPECS.	SPECIFICATIONS
MAX.	MAXIMUM	S.& P.	SHELF & POLE
M.B.	MACHINE BOLT	SQ.	SQUARE
M.C.	MEDICINE CABINET	S.S.	STAINLESS STEEL
MECH.	MECHANICAL	STD.	STANDARD
MIN.	MINIMUM	STL.	STEEL
M.O.	MASONRY OPENING	STOR.	STORAGE
MTD.	MOUNTED	SUSP.	SUSPENDED
MTL.	METAL	T.	TREAD
NAT.	NATURAL	T.C.	TRASH ENCLOSURE
N.I.C.	NOT IN THIS CONTRACT	T.& G.	TONGUE & GROOVE
NO.	NUMBER	THSLD.	THRESHOLD
N.T.S.	NOT TO SCALE	T.O.C.	TOP OF CONCRETE
OBS.	OBSCURED	T.O.M.	TOP OF MASONRY
O.C.	ON CENTER	T.O.W.	TOP OF WALL
O.D.	OUTSIDE DIAMETER	TYP.	TYPICAL
OPNG.	OPENING	U.N.O.	UNLESS NOTED OTHERWISE
P.L.	PLATE LINE	VERT.	VERTICAL
PLAS.	PLASTER	W.	WASHER
PR.	PAIR	W/	WITH
P.S.I.	POUNDS PER SQ. INCH	W/O	WITHOUT
P.T.	PRESSURE TREATED	WD.	WOOD
Q.T.	QUARRY TILE	W.H.	WATER HEATER
R.	RISER	WIND.	WINDOW
RAD.	RADIUS	W.P.	WATERPROOFING
R.D.	ROOF DRAIN	W.W.M.	WELDED WIRE MESH

D-3 GENERAL NOTES FOR A TYPICAL SET OF DRAWINGS

With our drawings we include a sheet titled "STANDARD NOTES AND DETAILS." It contains general notes, typical Building Department Requirements and standard details.

These **general notes** afford the owner a certain amount of legal protection and therefore should be included with each set of drawings, as mentioned in Chapter A-16.

GENERAL NOTES:

1. All construction to be in compliance with 1979 EDITION, Uniform Building Code and all other governing codes, ordinances and Fire Marshal regulations.

2. Contractor shall verify all dimensions, elevations, site conditions, including existing utility services, before starting work and the architect shall be notified immediately of any discrepancies.

3. Neither the owner nor the architect will enforce safety measures or regulations, they are the contractor's responsibility.

4. Contractor shall, as part of this contract, furnish all insurance, materials, labor, tools and equipment and services required to properly execute and complete his work according to the Drawings, Specifications, Addenda and Change Orders as issued.

5. Contractor is responsible for the finishing of his work in the manner and form prescribed by the Drawings and Specifications, but shall not be required to begin any part of the job where proper preparation preceding his work is not in order or complete. Any discrepancies shall be reported to the architect.

6. Contractor shall protect all property and the work of all other trades against damage or injury caused by his activity.

7. General contractor and subcontractors shall individually warrant for one year all materials and workmanship except as otherwise agreed.

8. In case of conflict, notes and specific details on drawings shall take precedence over these "General Notes" and over standard details. Where no construction details are shown or noted for any part of the work, details shall be the same as for other similar work.

9. Items specified on Drawings and Specifications represent the type and quality required. Con-

tractor and subcontractors may substitute "equal" items in their bids when approved by architect.

10. Contractor shall upon completion of his work, clean and clear the area of all debris or any other matter caused by his operation.

11. The owner reserves the right to change, increase or reduce the work as may be necessary and in such event shall notify the contractor in writing.

12. Contractor shall make no changes and do no extra work without written authorization from the owner or his representative.

13. The "General Conditions of the Contract for Construction" A.I.A. Document A-201, dated August 1976 are herewith made part of these notes. Copies of the document are on file with the owner, architect and general contractor.

P.S. The information above is included in this book to give you an idea of the type of general notes that should be a part of any set of drawings. However, don't just copy them verbatim; let your architect decide which ones apply. If you prepare the drawings yourself, use common sense to select the appropriate notes. Be sure the dates inserted with notes #1 and #13 are those of the latest editions of the UBC and A.I.A. 201, respectively.

D-4 USEFUL INFORMATION FOUND ELSEWHERE

What is it about?	Title of Publication and approximate cost	Where available
List of firms, products, trade names	Sweets Catalog	architects, contractors
Construction regulations (most Western states)	Uniform Building Code (UBC) ($40)	A.I.A. Chapter offices; technical bookstores; some Bldg. Departments
Excerpts from the UBC relating to houses and duplexes	Dwelling Construction under the UBC, one and two story dwellings ($8.50)	A.I.A. Chapter offices; technical bookstores; some Bldg. Departments
General building construction procedures	A.I.A. General Conditions ($1.50)	A.I.A. Chapter offices; architects, contractors, some blueprinting companies
Building and the law	Legal pitfalls in architecture, engineering and building construction	McGraw-Hill Book Co. 1221 Ave. of the Americas New York, NY 10020
Laws affecting contractors, lien laws	California License Law and Reference Books ($6.00)	General Services Publication Section P.O. Box 1015 North Highland, CA 95660
Legal Guide	Practical legal advice for builders and contractors by Edward E. Colby ($15)	Prentice-Hall, Incl Englewood Cliffs, NJ
Home building, practical advice to home building contractor. Of value to the layperson.	Home Builder's Guide by James D. Higson ($8)	Craftsman Book Company 6058 Corte Del Cedro Carlsbad, CA 92008
Understanding building construction drawings	Reading Construction Drawings by Paul I. Wallach and Donald E. Hepler ($29.95)	McGraw-Hill Book Co. P.O. Box 400 Hightstown, NJ 08520
Buying and building for rehabilitation	Rehabbing for Profit ($29.95)	McGraw-Hill Book Co. P.O. Box 400 Hightstown, NJ 08520

What is it about?	Title of Publication and approximate cost	Where available
Government Loan Programs	Builder's Guide to Government Loans ($13.75)	Craftsman Book Co. 6058 Corte Del Cedro Carlsbad, CA 92008
Financing of a commercial building	How to Get the Best Financing for Your Building (free)	VARCO-PRUDEN Bldgs. AMCA International Clark Tower, Memphis, Tennessee 38137
Residential construction costs	Estimating Home Building Costs ($14)	Craftsman Book Co. 6058 Corte Del Cedro Carlsbad, CA 92008
Residential framing and other wood construction	Wood Frame House Construction ($8.25)	Craftsman Book Co. 6058 Corte Del Cedro Carlsbad, CA 92008
Outstanding textbook on architecture and drafting	Architecture - Drafting and Design by Donald E. Hepler and Paul I. Wallach	McGraw-Hill Book Co. P.O. Box 400 Hightstown, NJ 08520
Loan costs and related expenses	Homebuyers' Guide to Settlement Costs (free)	HUD, available from most lending institutions
Log Houses, with plans, details and dealer references	Complete Guide to Buying, Building and Maintaining Log Homes ($10)	Home Buyer Publications 3534 Devon Drive P.O. Box 2078 Falls Church, VA 22042
Building with F.H.A.	Questions and answers on F.H.A. Home Property Appraisals (free)	See telephone book under United States Government Housing and Urban Development, Federal Housing Administration
Building with V.A.	Pointers to the Veteran Home Owner (free)	See telephone book under United States Government Veterans Administration
To buy or to rent? Financing, insurance	Selecting and Financing a home ($1)	U.S. Dept. of Agriculture Consumer Information Center, Dept. 091-F, Pueblo, CO 81009
What is and how to "open" escrow	Escrow (free)	Most construction lenders

What is it about?	Title of Publication and approximate cost	Where available
How Fund Control works	Fund Control (free)	Lumberyards, lenders
Building or remodeling a church	Branch Church Building (free)	First Church of Christ, Scientist, Boston, MA
Home building and decorating, with extensive product information	Home Magazine ($12 per year) A Best Buy!	Home P.O. Box 10002 Des Moines, IO 50347-0002
Home building, with good details of construction	Fine Homebuilding (magazine) ($14 per year) Excellent Buy!	The Taunton Press 52 Church Hill Road Box 355 Newtown, CT 06470
Energy conservation	Nonresidential Design Manual ($10.60) Residential Design Manual ($4.24) (add $1 for postage)	Publications Unit Calif. Energy Commission 1111 Howe Ave. Sacramento, CA 95825
A store specializing in technical books, codes, forms	Ask for latest catalog of "Technical Books, Codes and Forms"	Building News Inc. P.O. Box 3031 Terminal Annex Los Angeles, CA 90051 (213) 870-9871

INDEX

(Bold page numbers indicate definitions in the glossary D-1)

ORDER FORM

Please send me __ copies of the work manual HOW TO GET IT BUILT — better, faster, for less
[9] $20.00 per copy, sales tax and postage included. Full refund guaranteed if not satisfied, when book
is returned in good condition within 4 weeks.

Our shipments are usually made within 10 days.

☐ Check enclosed   Card No._____

☐ Mastercharge   Exp. Date_____

☐ BankAmericard/Visa   Signature_____

Print Name_____

Print Address_____

City_____ State_____ Zip_____

Mail to: Werner R. Hashagen & Associates, Architects
7480 La Jolla Boulevard, La Jolla, CA 92037
(619) 459-0122

ORDER FORM

Please send me __ copies of the work manual HOW TO GET IT BUILT — better, faster, for less
[9] $20.00 per copy, sales tax and postage included. Full refund guaranteed if not satisfied, when book
is returned in good condition within 4 weeks.

Our shipments are usually made within 10 days.

☐ Check enclosed   Card No._____

☐ Mastercharge   Exp. Date_____

☐ BankAmericard/Visa   Signature_____

Print Name_____

Print Address_____

City_____ State_____ Zip_____

Mail to: Werner R. Hashagen & Associates, Architects
7480 La Jolla Boulevard, La Jolla, CA 92037
(619) 459-0122

ORDER FORM

Please send me __ copies of the work manual HOW TO GET IT BUILT — better, faster, for less
[9] $20.00 per copy, sales tax and postage included. Full refund guaranteed if not satisfied, when book
is returned in good condition within 4 weeks.

Our shipments are usually made within 10 days.

☐ Check enclosed   Card No._____

☐ Mastercharge   Exp. Date_____

☐ BankAmericard/Visa   Signature_____

Print Name_____

Print Address_____

City_____ State_____ Zip_____

Mail to: Werner R. Hashagen & Associates, Architects
7480 La Jolla Boulevard, La Jolla, CA 92037
(619) 459-0122